VIDEOCASSETTE RECORDERS

VIDEOCASSETTE RECORDERS

A servicing guide
Third Edition

Steve Beeching, AMIERE,
TEng.(CEI), MRTS

Heinemann Professional Publishing

Heinemann Professional Publishing Ltd
Halley Court, Jordan Hill, Oxford OX2 8EJ

OXFORD   LONDON   MELBOURNE   AUCKLAND

First published by Butterworth & Co. (Publishers) Ltd 1983
Second edition 1985
First published by Heinemann Professional Publishing Ltd
Third edition 1988

British Library Cataloguing in Publication Data
Beeching, Steve
  Videocassette recorders : a servicing
  guide. – 3rd ed.
  1. Video tape recorders and recording –
  Maintenance and repair
  I. Title   II. Beeching, Steve.
  Videocassette recorders
  778.59′93′0288   TK6655.V5
  ISBN 0 434 90123 7

Photoset by Butterworths Litho Preparation Department
Printed in England by Richard Clay Ltd, Chichester, Sussex

CONTENTS

PREFACE

The world of video changes very quickly and what is more the rate of change is also increasing. Technical advancements made over the last five years will be complemented by changes that will be made over the next three years. Since the first edition of this book not only have the original Philips N series videorecorders become obsolete but also the V2000 series and Betamax in certain parts of Europe. This does not mean that the techniques described in this book are also obsolete where they refer to these systems as in time they will be resurrected in other forms as development continues. Many parts of videorecorder circuits are being integrated and often the repairer is faced with a large scale integrated circuit and no idea of what is happening inside it. I hope that the information contained within these pages will supplement the manufacturer's service manual by dealing with the techniques used in some detail. Manufacturers' service manuals on recent or later versions of models often do not contain the original circuit descriptions or information on VCR technology, only reference to the circuit techniques particular to the recent model. The role of this book is to provide a ready reference of all of that information.

I have included some information on the Video 8 system where it is relevant to video but not the PCM audio. S-VHS is described as a new format for the future to accommodate the very high quality pictures from high definition TV and satellite and for the home video enthusiast who wants near broadcast picture quality.

Up to date additions to the colour section and FM recording noise reduction systems as applied in HQ VHS are included to illustrate some of the more complex noise reduction systems employed in the higher grades of videorecorders. These will form the basis for technological advancements that will be found in future S-VHS recorders.

A large part of the video market is taken by camera/recorder combinations, referred to as camcorders, and I have added the VHS-C format and control of a four-head video drum found not only within C format camcorders but also full size camcorders.

Cameras form the front half of a camcorder and so the technology behind camera tubes and their colour processing have been added along with the very complicated world of the colour CCD pick up device, although timing and processing control of the solid state is all LSI. In the camcorder section I have also included the techniques used for autofocus cameras and camcorders.

Chapter 7 has been expanded to update the still picture and slow motion technology and includes the use of a digital field store. The use of digital technology is part of the rapid advancement of videorecorders to include picture within picture in future models by video signal digitisation. The system control circuits are now looking more like microcomputers but they still need the analogue interfaces to analyse the external sensors. Looking into the future I can foresee the increasing use of digital technology for audio, slow motion, automatic tracking and timebase correction built into the higher grades of videorecorders . . . it was once said 'you ain't seen nothing yet!'.

Steve Beeching

INTRODUCTION

The video recorder is split into seven main sections, of which only one is common to television receivers – the tuner/IF stage:

1. Tuner and IF stage.
2. Record and replay, monochrome, or luminance.
3. Record and replay, colour.
4. Servos.
5. Systems control and timer.
6. Power supplies.
7. Audio record and replay.

It is not intended to cover the tuner and IF stage, nor the UHF modulator, the former being a standard signal handling item and the latter a 'black box'. Power supplies are not covered either as these also tend to be fairly straightforward, and are covered by the manufacturers' service manuals. Servos, the colour signals and record/replay are considered in detail; 'linear' audio recording is a standard audio technique and is not covered; Hi-fi audio has a chapter to itself. Systems control is treated in general as individual manufacturers use 'in house' or 'dedicated' microprocessors for this purpose and detailed analysis of each one would be too much to cover at this stage; however certain techniques are covered in detail as examples.

A chapter is devoted to trick facilities, slow motion, still picture and fast search, showing individual approaches to the solution of the various problems that arise when trick facilities are incorporated.

The book is written for colour TV engineers and those taking Radio and Television Techniques courses. It is assumed that the reader has a basic knowledge of colour television and some experience in video recorders, even if this is only to know how to operate them.

Basic concepts

Consider an audio tape recorder which has a frequency response of 20 Hz–20 kHz and that it is desired to increase the range up to 3–5 MHz.

The first approach would be to increase the speed of the tape to obtain a higher frequency response from the head and the tape. The magnetic gap of the tape head would have to be reduced considerably. What sort of speed would be adequate? Rough calculations put us up to the region of 8 m/s, resulting in very short recording times or very long reels of tape, up to one metre in diameter. If we were able to solve the massive tape handling problem there would still be the problem of frequency response.

The amount of magnetic flux passed by a tape head to the tape is directly related to the current flowing in the windings, i.e. $I = V/Z$. As the frequency increases so does the head impedance, Z. In order to maintain the same value of recording current the drive coltage V, would have to be increased proportionately. This is called the pre-emphasis or recording correction, and V rises at a rate of 6 dB/octave. An increase of 6 dB applied to a voltage change is ×2 and an octave is an increase in frequency of ×2. So every time the frequency is doubled so is the voltage drive to the head.

If as an example the impedance $Z = 100\,R$ at 25 Hz, then at 25 kHz it would be 1000 times higher, that is 100 K. The recording drive voltage would have to be 1000 times more, an increase of 40 dB. Now if this is taken to 2.5 MHz the recording voltage would be about 1 kV, which is not very practical! Even by applying audio techniques to linearise the response the range required would still be too great, being 11 octaves.

Solving the two problems outlined over the years has resulted in the two fundamental video recorder techniques – helical scan and FM recording. Helical scan was developed by Toshiba in the years from 1954 to 1959 when the first machine was produced.

Helical scan

To obtain the required high speed of tape at 8 m/s it was found that it was possible to run the tape at a slower speed and then run the video heads at high speed thus maintaining high head to tape speed.

The tape travels around a revolving video head drum which is tilted at an angle to the tape. The video heads are very small and record magnetic video tracks, which, as will be seen later in much more detail, are recorded as slanting stripes on the tape. The head to tape speed is between 4 m/s and 8 m/s depending on the system. Whilst helical scan solved the high head to tape speed problem it created problems of its own, i.e. how to control the rotation of the video drum and to maintain continuity of the replayed signal, that is replaying without any sudden gaps. Servos are used to tightly control the tape speed and drum rotation in speed – to control the rotating drum at a constant speed (1500 rev/min) and for phase to control the rotation within one revolution at that speed of 1500 rev/min.

FM recording

One of the problems in helical scan is head to tape contact. Video heads can 'bounce' off the tape as it travels past the drum and through the mechanics, and another problem is bandwidth or frequency response. The video signal is frequency modulated onto a carrier. Sync tip relates to the lowest frequency, around 3 MHz, while peak white relates to the highest frequency, around 4.2 MHz.

So instead of trying to record a signal of 25 Hz –2.5 MHz we have a bandwidth of 3 MHz–4.2 MHz, which can more easily be handled by high frequency video heads. Also, in frequency modulation it is the bandwidth of the modulating signal which creates sidebands. The lower sideband is limited to 1 MHz, to allow for a colour undercarrier to be slotted in below 1 MHz. The upper sideband is limited to about 5.5 MHz, and both limits are shaped by filters. The sidebands then extend the bandwidth to be recorded from 1 MHz to 5.5 MHz; to maintain the upper limit some pre-emphasis is required and with de-emphasis on replay an acceptable signal to noise ratio is obtained.

Another advantage of an FM carrier is that it is immune to a large degree from amplitude changes if it is amplified and limited. This characteristic eliminates video head bounce and compensates for low amplitude replay signal levels created by mechanical tape handling tolerances, tracking shift and interchangeability with other recorders. Tracking control is less critical than it would be if FM were not utilised.

1

SYSTEMS

The Philips N1500

A suitable starting point in any introduction to video tape recorders is the tape path, first developed on a cassette basis in the Philips N1500. To illustrate the requirements for satisfactory tape handling through the mechanics we shall go

through this system in detail and explain the function of the various guide posts.

The tape spools in the Philips N1500 system are mounted coaxially, one above the other. The lower spool is the supply spool and the upper one is the take up spool; Figure 1.1 shows the tape path. The tape leaves the supply spool inside the cassette housing and travels out of the cassette around an internal roller (K); a small amount of braking is applied to the supply spool for tape tension, but more of that later. The tape then passes over the full erase head so that tape to be recorded is wiped clean. The next guide (L) is a height adjustment guide having flanges top and bottom (see Figure 1.2) to determine the height adjustment of the tape passing the erase head and it also provides pre-adjustment for the tape path to the drum assembly to guides M, M' and N. M is an insulated pillar, it is not connected to chassis but L is, a small strip of metal foil tape connecting L and M together at the beginning and end of the tape will stop the tape drive. M' is a shaped pillar; the conical shape forces the tape downwards as it passes the guide, giving the tape a downward 'bias' so that when the tape has passed around the post N, which is also slightly conical, it sits down on the cam (O).

The cams O, P and R are ruler edge cams and as the tape is being biased downwards it sits on the

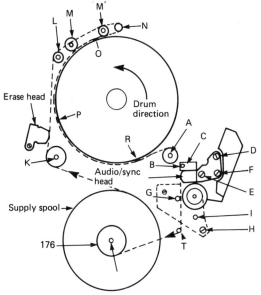

Figure 1.1.   Philips N1500 tape path

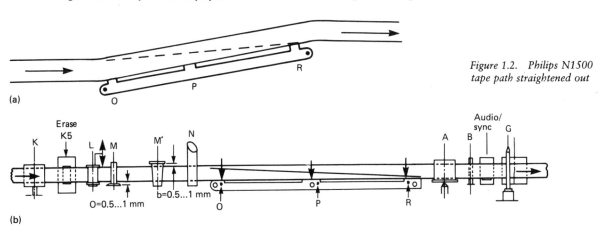

Figure 1.2.   Philips N1500 tape path straightened out

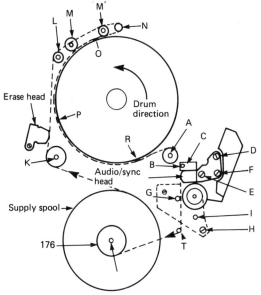

edges around the drum assembly and it should not wander up and down. After cam R the tape continues around roller (A), which is also an internal cassette housing roller, and then passes pillar (B), which stabilises the tape against any ripples on the edges caused by the transport mechanism which it has already gone through. The audio/sync head has the control track (sync) section at the top and the audio record/play section at the bottom (see track pattern data, Figure 1.4). After the audio/sync head the tape is held tight by the pinch wheel and capstan drive spindle from which the tape drive is provided. The tape speed through the mechanics is governed by the capstan which is servo controlled. The capstan is also part of the tape downward bias system; the capstan spindle is tilted slightly in the direction shown by the arrow in Figure 1.2 which puts a 'tilt' into the tape path by forcing the tape upwards as it passes the capstan which also forces it downwards before it reaches the capstan. Whilst this may be difficult to visualise it may be easier to treat the tape as a solid strip and not a flexible ribbon when the biasing becomes a bit easier to understand. Figure 1.2 shows the tape path stretched out so that the guides and the height adjustments can be seen more clearly. While Figure 1.2b shows the drum assembly cams O, P and R as a straight line they are in fact a slope, as shown in Figure 1.2a, to take into account the split level spools. The dotted line is the path of the video heads, and shows how the video tracks are recorded on slanted tracks with respect to the tape.

A certain amount of take up torque is applied to the take up spool and the tape is fed to this spool around a pillar (T) which is mounted within the cassette housing.

Figure 1.3 shows the concentric spool carrier assembly, the upper turntable carrying the take up spool and the lower turntable the supply spool. In play or record modes the upper turntable is driven via a felt ring clutch assembly (not shown) and the supply spool is 'free running' but sits on the felt ring shown. The felt ring is held still while the turntable runs on it, and the felt thus acts as a gentle brake and provides 'back tension', which should not be so great as to cause the tape to stretch or too loose so as to cause looping. The aim is to keep the tape taut but not tight so that it acts as if it were a solid strip and maintain pressure on the erase and audio/sync head surfaces for good contact, and for video head contact as well as the requirement of downward bias.

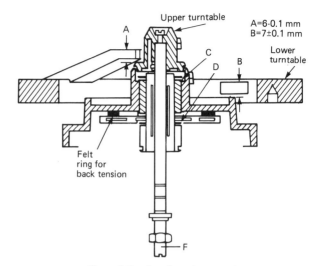

Figure 1.3. Spool carrier assembly

So much for the mechanical tape path, but how does this relate to the television signal that is to be recorded? Figures 1.4 and 1.4a show the magnetic tracks on the tape and the specification of the N1500 system.

As we have seen the tape travels in the same direction as the video heads. The heads contact the tape at the top edge and leave it at the bottom edge, thus recording a slanting track. Figure 1.4 illustrates the tape ribbon with the magnetic tracks on it.

Each edge of the tape is an audio track, although the upper one, track 2 is not used, except on the semi professional N1520. For general considerations it is better to define the sync or control track as it is sometimes called, as being the upper edge of the tape and the audio track as being on the lower edge of the tape. Using Figure 1.4 we can relate the video signal to the magnetic track, not forgetting that it is recorded on an FM carrier. Point 1 on the upper edge is the beginning of the magnetic track when the rotating video head contacts the tape, and is related to a point on the video signal about eightlines before the field sync pulse. Relative to the TV picture this point is at the bottom of the screen, and is not normally seen on a TV with standard height adjustment. Further along the track is point 2, the beginning of the field sync pulse and point 3 is the end of field blanking when video is present. Point 3 is at the top of the TV screen, while the remaining part of the track down to point 4 is the TV picture as viewed.

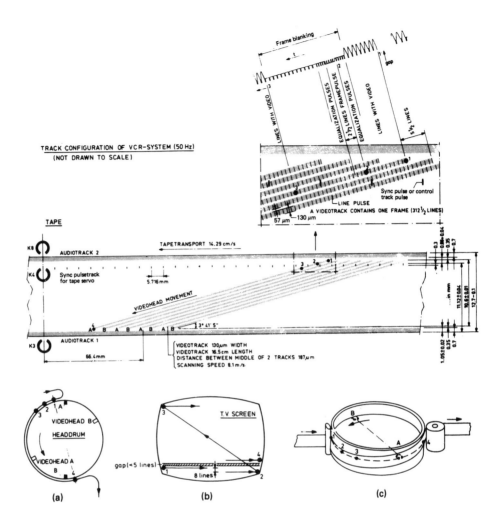

TRACK CONFIGURATION OF VCR–SYSTEM (50 Hz)
(NOT DRAWN TO SCALE)

A VIDEOTRACK CONTAINS ONE FRAME (312½ LINES)

TAPE

TAPETRANSPORT 14.29 cm/s

AUDIOTRACK 2

Sync pulsetrack
for tape servo

5.716 mm

VIDEOHEAD MOVEMENT

3° 41′ 5″

AUDIOTRACK 1

66.4 mm

VIDEOTRACK 130μm WIDTH
VIDEOTRACK 16.5cm LENGTH
DISTANCE BETWEEN MIDDLE OF 2 TRACKS 187μm
SCANNING SPEED 8.1 m/s.

57 μm — 130 μm

VIDEOHEAD B
HEADDRUM
VIDEOHEAD A

T.V SCREEN

gap(<5 lines)

8 lines

Figure 1.4.
Magnetic tracks of the
N1500 system

(a)   (b)   (c)

There is, however, an overlap of magnetic tracks related to the video signal. This is explained by the fact that point 4 at the end of the video track also contains some of the eight lines being recorded by the second video head at the beginning of its own magnetic video track. So at any time, when one video head is just finishing recording, the other head has already started to record. The eight to ten lines of video which occur prior to field syncs are in a brief period when both video heads are in contact with the tape.

The reason for the overlap of recorded information for the period of head changeover is so that when the FM carrier signal is replayed both video heads replay at the same time for just a few lines in order to prevent any gaps or discontinuity of the replayed FM signal.

To further support this topic, whilst we say that the tape is in contact with the video drum for half of its circumference, or 180° wrap around, the wrap around is in fact 186°, to provide for the small overlap period. This overlap period is called the head crossover point, in replay, the replay head crossover point, as opposed to the record crossover point. The precise crossover point is when head 1 stops replaying and head 2 takes over, the changeover in replay controlled by a switching signal developed from the drum servo. Crossover points can be clearly seen by reducing the height of a TV in replay, when a small thin horizontal line appears at the bottom of the picture, about five to ten lines before field blanking.

Figures 1.4a, b and c illustrate the magnetic tracks and points 1 to 4 as they occur with respect to the video heads and drum and on the TV picture. This will enable the reader to relate the slanting magnetic track on the tape to the actual picture and video recorder mechanics. From Figure 1.4a it can be seen that the tape wrap is slightly more than 180°, that is 186°, and that head A contacts the tape while head B is still in contact and about to leave. An obvious question that would be asked after studying Figure 1.4 is: 'Why do the control track pulses not interfere with the video track as they *are* placed within it?' The answer is that, if you look carefully at the magnetic track, the control track pulses are positioned between points 2 and 3 and therefore will be replayed by the video as a pulse or spike but during the field blanking period. Also note that there is one control track pulse for every two video tracks, a point which will be clarified in the chapter on servos. This of course means that the control track pulses are recorded at frame rate, not field rate and that one control track pulse occurs for every revolution of the drum.

Basic mechanical data
VCR system (N1500)

Drum diameter	105 mm
Speed, head disc	1500 rev/min
Scanning speed (relative speed, video head/tape)	8.1 m/s
Tape speed	14.29 cm/s
Video head gap length	0.8 μm
Video track width	130 μm
Distance between two video tracks	57 μm
Audio track width	0.7 mm
Sync track width	0.3 mm

The control track pulse replayed during the field blanking period did not usully cause any problems, but in the early years some TVs did not have sufficient blanking drive to the tube. The control track pulse was able to break through and could be seen as a dancing multicoloured spot on the picture. Later versions (N1501) of the video recorder had a spot suppression circuit added to provide additional field blanking to the replayed signal and thus eliminate the spot. The system parameters are listed in the table for comparison with later systems. The head to tape speed of 8.1 m/s should be noted.

Note also that each magnetic track is 130 μm wide (1 μm is one millionth of a metre) and that there exists a guard band or space between each magnetic track which is 57 μm wide. This spacing is there to prevent a video head, which may have wandered slightly off track during replay, from picking up any information from adjacent video tracks which would cause intereference on the FM carrier.

Philips N1700 Format

The N1700 system is basically a slowed down N1500 system, and there are discrete minor changes in the mechanics to achieve the required conditions for other parameters.

A comparison of tape speeds reveals that the N1700 at 6.56 cm/s is less than half of the N1500 speed which is 14.29 cm/s. The speed reduction and an increase in the length of tape in a cassette enabled the 1 hour time of the N1500 to be extended to the 2½ hours of the N1700. Note though that the head to tape speed is unchanged at 8.1 m/s.

No guard band is present on the N1700 and the magnetic tracks are reduced in width from the 130 μm in the N1500 to only 85 μm in the N1700; however this is still wider than the VHS (49 μm) or the Betamax (32.8 μm) and the V2000 at 22.5 μm. Each new system is reducing the physical parameters to obtain longer and longer playing times.

One minor point not normally considered in the N1700 is the change in mechanical track angle from 3°41'5" in the N1500 to 3°42'52". This change in track angle is to accommodate yet another new parameter, 1½ lines offset between magnetic tracks. Also the video heads mounted on the head drum are not 180° apart but 179°25'45", which is approximately one TV line displacement or offset. The result of the two minute changes gives rise to the 1½ lines offset between each magnetic video track with respect to the one before. The reasoning for this offset is important in the colour crosstalk elimination process for the N1700; it is to ensure that any two lines lying next to each other in two adjacent field tracks have the same PAL colour phase (see Chapter 5).

It does mean that modifications to N1500 that update them to N1700 or long play do not take into account this 1½ lines offset and so it is not accurate and colour crosstalk is evident.

Video head displacement

Displacement: 180° − 179°25'45" = 34'15" which, expressed in seconds, is 2055".

Now 360° is equivalent to 625 TV lines
   360° = 1 269 000"
   Therefore 1 269 000 ÷ 625 = 2030"

The displacement of the video heads is 2055", approximately equal to 1 TV line, which is 2030". This displacement of the video head gives a 3½

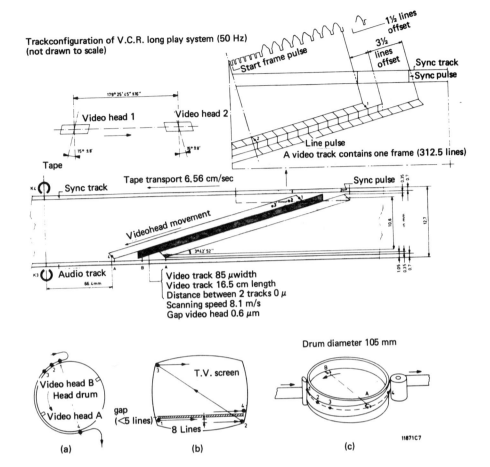

Trackconfiguration of V.C.R. long play system (50 Hz)
(not drawn to scale)

1½ lines offset

3½ lines offset

Start frame pulse

Sync track
Sync pulse

Video head 1   Video head 2

179° 25' 45" ±16"

15° ±5'   5° ±5'

Line pulse
A video track contains one frame (312.5 lines)

Tape

Sync track   Tape transport 6.56 cm/sec   Sync pulse

Videohead movement

3° 42' 52"

Audio track
66.4 mm

Video track 85 μwidth
Video track 16.5 cm length
Distance between 2 tracks 0 μ
Scanning speed 8.1 m/s
Gap video head 0.6 μm

Drum diameter 105 mm

Video head B
Head drum

Video head A

T.V. screen

gap
(<5 lines)

8 Lines

(a)   (b)   (c)

Figure 1.5.
Main parameters of the
N1700 system

line offset between Track A and Track B and a 1½ line offset between Track B and Track A. The aim is to provide a 1½ line shift between each successive recorded track (see also p. 61).

Figure 1.5 shows the main parameters. The video heads have an azimuth of +15° and −15° totalling 30°. The control track is on the top edge

of the tape and is not recorded within the bounds of the video tracks as it was in the N1500 format, consequently there is no problem of interference by the video heads replaying the control track.

It can be seen that there is no guard band and that magnetic field tracks lie adjacent to each other and the azimuth tilt between the video heads takes care of FM carrier crosstalk (see Chapter 2). The 1½ lines offset shown between each track helps to reduce the effects of colour crosstalk. The audio track is on the bottom edge of the tape and there is only provision for one audio signal and not two.

The VHS format

In the VHS system the cassette has two tape spools, take up and supply, mounted side by side inside a flat cassette housing. They are not stacked as in the earlier Philips system. The maximum playing time was 3 hours but this has been extended by the use of a thinner tape to 4 hours.

Basic mechanical data VCR system (N1700)	
Drum diameter	105 mm
Speed, head disc	1500 rev/min
Scanning speed (relative speed, video head/tape)	8.1 m/s
Tape speed	6.56 cm/s
Video head gap length	0.6 μm
Video track width	85 μm
Distance between two video tracks	0 μm
Audio track width	0.7 mm
Sync track width	0.7 mm

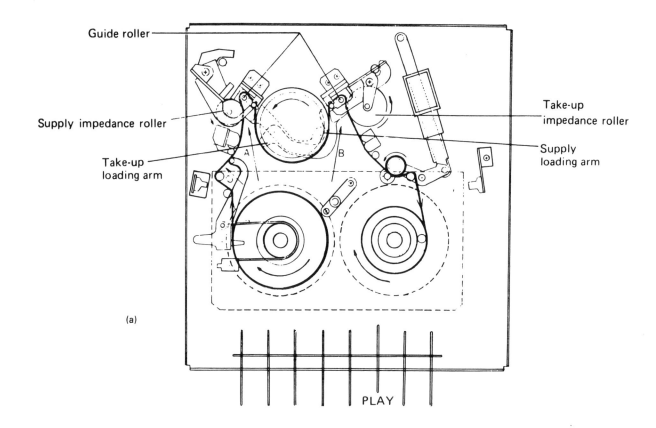

(a)

PLAY

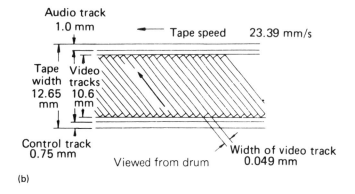

Audio track
1.0 mm

Tape speed   23.39 mm/s

Tape
width   Video
12.65   tracks
mm   10.6
         mm

Control track
0.75 mm

Width of video track
0.049 mm

Viewed from drum

(b)

Guide roller

Supply impedance roller

Take-up
loading arm

Take-up
impedance roller

Supply
loading arm

Figure 1.6.   a, the VHS parallel loading system.
b, the VHS video tape format

With the cassette inserted into the videorecorder as shown in Figure 1.6a, the supply spool is on the left hand side and the take up spool is on the right hand side. Mounted either side of the cassette compartment are the end of tape sensors, the left hand one detects the end of the tape and hence is the 'End Sensor'. The right hand one is the 'Start Sensor' which detects the beginning of the tape at the termination of rewind. End of tape detection is optical and a bulb is mounted in a central position between the spools, projecting inside the cassette package and aligning with the sensors through slots in the cassette housing. At each end of the tape is a clear leader section so that the bulb can illuminate the sensors through the clear leaders and stop the tape.

After the cassette has been inserted and play/record has been selected, threading takes place. Tape is drawn out of the cassette by loading arms, which are also the drum assembly entry and exit guides and loading is in the direction of the arrows A and B. The arms locate in specially aligned end stops and are clamped: Figure 1.6a illustrates the tape in the loaded position. The VHS system uses the 'M' wrap technique, so called since the tape path is similar to the letter M, also used by Philips in the V2000 system. In play mode the tape travels out of the supply spool, around internal cassette roller posts and then exits the cassette to the tension arm pillar.

The tension arm functions as a variable brake onto the supply spool, and depends upon the tension of the tape passing the tension arm to vary the braking effect upon the supply spool. This is mechanical negative feedback and it maintains constant tension upon the tape being drawn out of the supply spool irrespective of the amount of tape on the spool.

The tape passes the full track erase head after the tension arm, then passes over an impedance roller to stabilise the tape ribbon prior to entering the drum assembly via the entry guide. It is the function of the entry guide to 'bias' the tape in a downward direction to sit it upon the ruler edge around the video head drum. The ruler edge around the drum is the mechanical VHS track standard and the tape travels upon the edge. The exit guide also biases the tape onto the ruler edge as it leaves the drum assembly, ensuring full contact around the drum.

After the exit guide the tape passes over another impedance roller before making contact with the audio/sync head, the roller removing ripples which would cause fluctuation in sound levels. The cap-

stan drive spindle and pinch wheel provide constant tape speed and the tape is wound onto the take up spool under tension. The take up spool is driven by a slipping clutch mechanism to provide take up tension; in later models the clutch is replaced by a spool drive motor which is lightly driven by a low voltage supply.

Back tension is adjusted by a spring bias onto the tension arm and it is set to between 28 and 45 g cm. Take up tension is set by the force on the take up clutch to between 150 and 220 g cm.

Tape pattern

Tape enters the video drum assembly at a high level and exits at a low level, which enables the video head to contact the tape at the bottom edge and leave it at the top edge. To keep the tape level with respect to the cassette, the drum assembly is tilted slightly.

Figure 1.6b illustrates the magnetic track pattern as seen on the oxide side of the tape from the drum, the direction of the tape is right to left and the video head scans upwards as shown by the arrows. Tape speed is 2.339 cm/s and the video tracks are 49 μm wide. Head to tape writing speed is 4.86 m/s.

VHS has two video heads mounted 180° apart. Track angle and speed provide a 1½ line offset between the start of adjacent field tracks. Azimuth tilt technique is used for adjacent track FM crosstalk rejection, the tilt being +6° and −6° for each head.

The audio track is on the upper edge and the control track on the lower edge, with provision made within the VHS system for stereo audio.

Basic mechanical data VHS system	
Drum diameter	62 mm
Speed, head disc	1500 rev/min
Scanning speed (relative speed, video head/tape)	4.86 m/s
Tape speed	2.339 cm/s
Video head gap length	0.3 μm
Video track width	49 μm
Distance between two video tracks	0 μm
Audio track width	1.0 mm
Sync track width	0.75 mm

Betamax system

The Betamax system, like VHS, also has two parallel spools mounted side by side. The cassette package is smaller than VHS, and it carries less tape; the maximum record/play time is 3 h 40 m at the time of writing and an extension is expected.

The tape loading is similar to the professional U-matic, although loading direction is anticlockwise whereas in U-matic it is clockwise. Figure 1.7 illustrates the Beta loading system. This tape path

erase head. The mounting assembly of the erase head consists of ceramic plates top and bottom forming the entry guides, to 'bias' the tape ribbon downwards onto the drum cylinder ruler edge.

Tape travels around the drum on a 186° wrap around, and leaves the drum assembly via the audio and control track heads, which are also mounted on ceramic plates forming the exit guides. Note that the Grundig/Philips system does *not* have a control track head. Tape then travels between the capstan drive spindle and the pinch

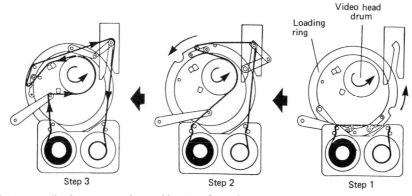

Figure 1.7.   The Betamax loading system, also used by Grundig in the V2000, and by Philips in the V2025 and later models

is also used by Grundig in the V2000 format, and Philips have also replaced their 'M' wrap format with the Beta wrap in models starting from the V2025.

When loading takes place the loading ring rotates anticlockwise as the small loading arm, also containing the pinch wheel, pulls the tape out of the cassette, Step 1. In Step 2 it is about two thirds of the way through the loading process, and a guide has pivoted into the right hand corner to keep the tape return path away from the drum. Step 3 shows loading completed. The main components of the deck assembly are shown in Figure 1.8.

The tape leaves the supply spool via internal guide posts and travels around the tape tension regulator. Tension is regulated in the same manner as previously described by mechanical feedback to the supply spool brake.

Grundig V2000 uses two motors for take up and supply, and cassette spool turntables are mounted directly onto the motor shaft. Back tension is applied to the supply spool motor as a light reverse drive and the tension arm is used as part of the sensor to control the motor drive, thus forming an electronic/mechanical feedback loop.

After the tension regulator arm, the tape enters the video drum assembly, passing the full track

wheel, which is the tape drive, around the tape extraction guide post and peripheral guide posts before entering the cassette package. As in all systems, the take up spool is also lightly driven to provide take up tension.

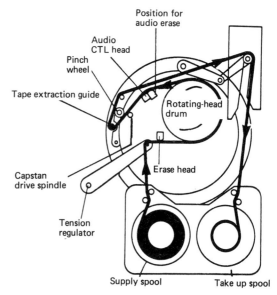

Figure 1.8.   Main components of the deck assembly

Beta tape format

The main components are shown in Figure 1.9. Entry guides forming the erase head assembly are shown. An assembly above the drum has two small springs with ceramic tips to bias the tape ribbon down as it travels around the drum. Note

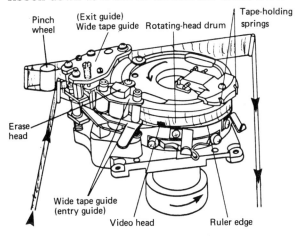

Figure 1.9. View of the Betamax tape wrap

that the video head tip contacts the tape at the lower edge, and that it leaves the tape at the upper edge.

The Betamax format is shown in Figure 1.10. Tape speed is 1.873 cm/s and the writing speed is 5.832 m/s. Video tracks start on the bottom edge and finish at the top edge and the head travels in the same direction as the tape, on tracks that are 32.8 μm wide. The audio track is on the top edge and the control track on the bottom edge and the tape is 12.65 mm (½ in) wide.

Azimuth slant technique is used as the field tracks are adjacent, so that FM crosstalk is reduced by the azimuth slant being $\pm 7°$. Also shown is the magnetic track angle (α) for moving tape which is $5°00'58''$. The stationary angle is in fact the angle of the ruler edge around the drum with respect to the deck mechanics and this is given as $5°00'00''$. This means that in a still picture mode the offset or error between replay track angle and recorded track angle is $58''$, which is too much with 32 μm tracks to enable a noise free still picture.

Tape end sensors in the Beta system are inductive; each sensor forms part of an oscillator and is a tuned circuit. At each end of the tape are metallic strips which, when they come into contact with the inductor, cause it to saturate. A saturated inductor has a very low impedance and oscillation cannot be maintained; a detector circuit monitors the oscillations and a 'stop' signal is generated.

Basic mechanical data Betamax system	
Drum diameter	74.487 mm
Speed, head disc	1500 rev/min
Scanning speed (relative speed, video head/tape)	5.832 m/s
Tape speed	1.873 cm/s
Video head gap length	0.4 μm
Video track width	32.8 μm
Distance between two video tracks	0 μm
Audio track width	1.05 mm
Sync track width	0.6 mm

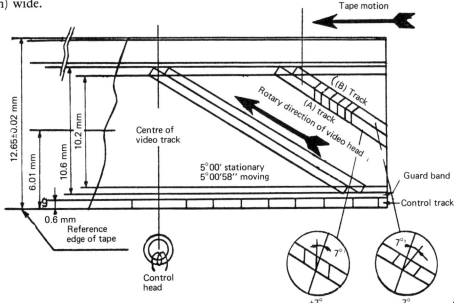

Figure 1.10.
The Betamax tape format

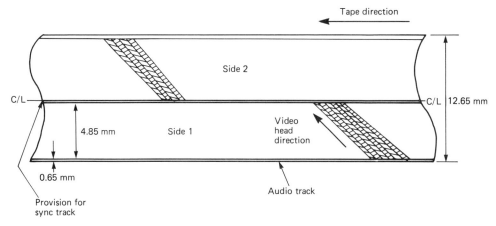

Figure 1.11. The V2000 tape format

V2000 system

In the V2000 system two tape wraps were initially adopted. Philips had an 'M' wrap system, similar to the VHS format, but slightly larger. Grundig adopted the Beta wrap mechanics which have proved more effective, but the format of the magnetic recording is very different. The end of tape stops on the Grundig V2000 series are reflective metallic strips which pass over an optocoupler and phototransistor arrangement.

Figure 1.11 is the basic recording format as seen from the video drum side of the magnetic tracks. When the tape passes through the drum assembly the active area is the lower half of the tape. The video heads contact the tape on the bottom edge, after the audio track, and leave it at the centre of the tape. The active video tracks cover a tape width of 4.85 mm and are 22.5 μm wide so that the V2000 system is effectively ¼ in tape.

Either side of the video track area are two tracks; the lower one is the audio track, 0.65 mm wide, the upper one is almost along the centre of the tape ribbon and is about 0.4 mm wide, designated control track. In fact the control track is not used in the V2000 system by Grundig. Philips use it for the 'GO TO' programme reference system in the VR2020 models, as a counter track.

This control track could also be used as a second audio track if required. It is difficult to visualise the principles of a video recorder within the V2000 system having auto reverse without requiring a second video head drum. The tape format is duplicated on the upper half of the tape and is used when the cassette is turned over. The

cassette package is a turnover cassette, as both 'sides' can be recorded and replayed.

Cassette packages are designed to 'instruct' the video recorder as to the cassette length, one hour per side, two hours per side etc., up to four hours per side. The longest is four hours per side or 'VCC 480' totalling eight hours of available recording time within a single package.

Basic mechanical data V2000 system	
Drum diameter	65 mm
Speed, head disc	1500 rev/min
Scanning speed (relative speed, video head/tape)	5.08 m/s
Tape speed	2.442 cm/s
Video head gap length	0.2–0.3 μm
Video track width	22.5 μm
Distance between two video tracks	0
Audio track width	0.65 mm
Sync track width (cue track)	0.4 mm
Video head width	23 μm

Fundamentals of V2000 automatic tracking

Dynamic Track Following (DTF)

The frequency spectrum of the V2000 system is shown in Figure 1.12. The FM recording spectrum is modulated between 3.3 MHz (sync tips) and 4.8 MHz (peak white) with sidebands. Colour under carrier (see Chapter 5, p. 73) is centred on 625 kHz.

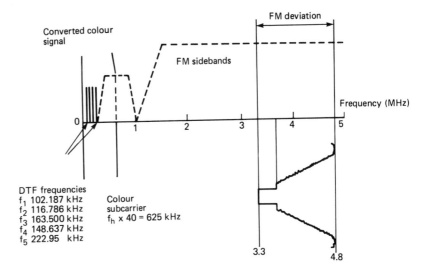

Figure 1.12. Frequency distribution in the Video 2 × 4 system

DTF frequencies
f_1 102.187 kHz
f_2 116.786 kHz
f_3 163.500 kHz
f_4 148.637 kHz
f_5 222.95 kHz

Colour
subcarrier
f_h × 40 = 625 kHz

In an area below the lower sideband of the colour under carrier five other frequencies are situated which are designated F1 to F5, F1 to F5 are the Dynamic Track Following frequencies.

Precise frequencies	Rounded off
F1 102.187 kHz	102 kHz
F2 116.786 kHz	117 kHz
F3 148.637 kHz	149 kHz
F4 163.500 kHz	164 kHz
F5 222.950 kHz	223 kHz

Record mode

In the record mode the DTF frequencies are recorded in the sequence F1, F2, F4, F3, as in Figure 1.13. Video head Ch.1 records F1 and F4, whereas Ch.2 records F3 and F2. The sequence is more obvious in the replay mode. The most important function for F5 is record, when it is used to space the 22.5 μm wide video tracks 22.5 μm apart, that is of course next to each other.

In the V2000 system the video heads are mounted on piezo-electric crystals called actuators. By driving the piezo crystals from a push-pull amplifier with a supply voltage of +150 V and −150 V it is possible to 'bend' the video heads up or down in the vertical plane.

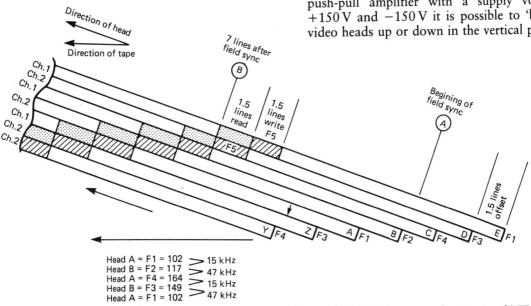

Head A = F1 = 102 ⟩ 15 kHz
Head B = F2 = 117 ⟩ 47 kHz
Head A = F4 = 164 ⟩ 15 kHz
Head B = F3 = 149 ⟩ 47 kHz
Head A = F1 = 102

F5 = 223 kHz

Figure 1.13. V2000 system track patterning of DTF frequencies

11

In record, actuator 2, upon which video head Ch.2 is mounted, is fixed to a reference voltage whereas actuator 1, upon which video head Ch.1 is mounted, is variable.

In order to generate a signal for actuator 1 in record each video head is fed with 223 kHz, F5, for a period of 1½ lines, 7 lines after field syncs, this is a 'write' period. Immediately after the 'write' period is a 1½ line 'read' period. Refer to Figure 1.13 and take for a starting point field track A which is recording with video head Ch.1 and with F1, DTF. F1 is recorded along with the video FM signal, and colour under carrier up to the 1½ line 'write' period. Then F5 is recorded instead of F1, but for only the 'write' period. This is then followed by the 1½ line 'read' period, and during this period the level of F5 from the previous track is read and stored.

The next field track, B, for Ch.2 video operates in the same way and F5 from track A is read during the 'read' period of B.

Assume that the first reading during track A is higher than the second sample in track B. This would mean that Ch.1 video head is low in height as indicated by the small arrow in Figure 1.13 and hence a correction voltage would be applied to actuator 1 to raise Ch.1 video head.

As actuator 2 is fixed the action of the record sensing circuits is to space Ch.1 track recordings equally between successive Ch.2 track recordings.

DTF frequencies replay

In replay, the DTF frequencies play a major role in tracking and controlling the capstan servo. The basis is the replay of a DTF frequency from a given video track along with crosstalk from the two adjacent video tracks, either side of it.

The total DTF replay is fed via a low pass filter to a mixer (Figure 1.14). A DTF frequency is generated from the pulse module and also fed into the mixer which will be the same sequences as they were in record (F1, F2, F4, F3).

Two filters are connected to the output of the mixer, 46/47 kHz and 15 kHz, and the outputs of the filter are buffered by amplifiers prior to mixing in a differential amplifier.

We will take as our example track A in Figure 1.13. The DTF frequency for this track is 102 kHz, DTF frequencies in track B and Z either side are F2, 117 kHz and F3, 149 kHz. If the video head replaying track A is 'on track' then equal levels of the DTF crosstalk frequencies 117 kHz and 149 kHz will be reproduced. The mixer will add and subtract all three DTF frequencies and two outputs will result from the filters: 117 kHz–102 kHz, i.e. 15 kHz and 149 kHz–102 kHz, i.e. 47 kHz. The output levels of these difference frequencies will be equal if the crosstalk frequencies replayed are equal, and the following differential amplifier will balance. If video head Ch.1 in replay

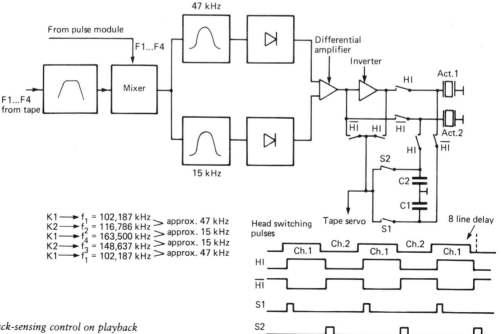

$K1 \rightarrow f_1 = 102,187$ kHz $\rangle$ approx. 47 kHz
$K2 \rightarrow f_2 = 116,786$ kHz $\rangle$ approx. 15 kHz
$K1 \rightarrow f_4 = 163,500$ kHz $\rangle$ approx. 15 kHz
$K2 \rightarrow f_3 = 148,637$ kHz $\rangle$ approx. 15 kHz
$K1 \rightarrow f_1 = 102,187$ kHz $\rangle$ approx. 47 kHz

Figure 1.14.   Track-sensing control on playback

12

is too low, again as indicated by the arrow, then the crosstalk outputs will be different. The 149 kHz output will be higher than the 117 kHz, therefore the outputs from the filters will be different, the 47 kHz output will be higher. An increase in level to the differential amplifier (Figure 1.14) will result in a fall at the output, then, via an inverter, an *increase* in voltage to actuator 1 which will then correct Ch.1 video head.

The inverse occurs if the video head is high, there is an increase in the output from the 15 kHz filter which will result in Ch.1 video head receiving a negative voltage and consequently the actuator will lower the video head. For Ch.1 video head, if it is low then an increase of 47 kHz results, whereas for Ch.2 video head the same situation will result in an increase in the 15 kHz output. To correct for this, switching pulses HI and $\overline{\text{HI}}$ switch actuator 2 to the output of the differential amplifier resulting in a correcting inversion. Timing pulses are shown in Figure 1.14. DTF switching pulses HI and $\overline{\text{HI}}$ are delayed by eight lines after the head crossover point.

In order to maintain the actuator drive level to a video head whilst it is *not* scanning a further switching arrangement is used. This eliminates the possibility of actuator drifting and then requiring correction upon commencing scanning again.

When Ch.1 video head is scanning switches HI in Figure 1.14 are closed and for a brief period S1 closes; this charges C1 up to the control voltage present for actuator 1. A short time later Ch.2 video head is scanning and switches $\overline{\text{HI}}$ are closed. One of these switches connects capacitor C1 to actuator 1, maintaining the stored control voltage until scanning commences again. There is an output from the DTF circuits to the tape, or capstan servo, during replay only, to accelerate or decelerate the tape drive if the DTF correction voltages

are too high. For example if both actuators are symmetrically correcting positive and negative during normal tracking correction the effect on the capstan servo after integration is minimal.

On the other hand, if both actuators require consistent negative correction then the integrated voltage is negative which will then accelerate the tape to compensate. In cases when either the DTF automatic tracking or the capstan servo circuits are faulty then both systems are erratic. It is easier for fault finding to check the capstan servo in record mode. If it functions correctly then the fault lies in the DTF circuits.

In earlier versions of the V2000, i.e. 2 × 4, the system is analogue. In the later 2 × 4 Plus, and 2 × 4 Super, control is microprocessor based and is digital with analogue to digital and digital to analogue converters.

The DTF automatic system allows for slow motion replay and still picture replay without tracking errors. This is extended in the 2 × 4 Super to include 7× fast forward, cue, and 5× reverse, review picture search modes without tracking errors. See Chapter 7 for details.

8 mm format

8 millimetre videomovie is a nomenclature for a video format which is derived from the tape width and is marketed towards replacing 8 mm cine film. It was heralded as a new technology format to replace all existing video formats, including VHS. However, this has not yet been proved to be so.

Nevertheless, there are advantages in specifying a new video format as the latest technologies can be incorporated: for instance, the 8 mm video system accommodates linear, FM and pulse code modulated (PCM) audio, as well as automatic tracking.

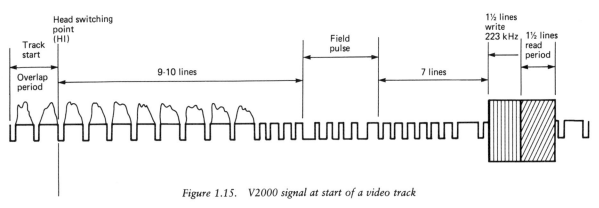

Figure 1.15. V2000 signal at start of a video track

Drum diameter	40 mm
Speed, head disc	1500 rev/min
Scanning speed (relative speed, video head/tape)	3.12 m/s
Tape speed	2.0051 cm/s (1.0058 cm/s LP)
Video head gap length	0.2 μm
Video track width	34.4 μm (17.2 μm LP)
Distance between two video tracks	0
Audio track width (auxiliary)	0.5 mm
Sync track width (cue track)	0.5 mm

The magnetic tape format is shown in Figure 1.16 with the cue track on the top edge of the tape, the auxiliary audio track on the bottom edge, with the video and PCM tracks in between.

The total tape wrap around (Figure 1.17) is 221°. The video tracks are recorded in 180° and follow a normal video technique, with the video heads taking 20 ms to record/replay because the drum is rotating at 1500 rpm which is equivalent to 40 ms per revolution — as in VHS or Betamax. In 8 mm format, however, the band of tape used for video tracks is only 5.351 mm wide.

An extension of the video track is used for PCM audio which is recorded and replayed by the video heads in an extra 31° of wrap around before the video tracks. The overlap margins at the start and finish of head scan of 5° each take up the remaining 10° of the total wrap around of 221°. Automatic track following (ATF) tones are also recorded on the video tracks, as described later.

The two video heads have a thickness of 27 μm with an azimuth tilt of ± 10°. Standard play video tracks are 34.4 μm wide but with video heads of

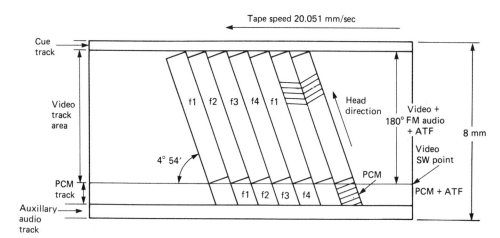

Figure 1.16.   8 mm video tape format

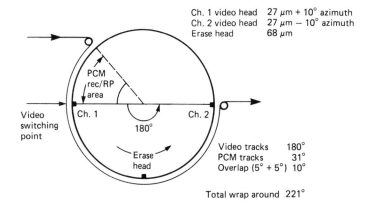

Figure 1.17.   8 mm video wrap around

only 27 μm there is an effective 7.4 μm wide guard band gap between each video track. In long play mode the video tracks are only 17.2 μm and this is obtained by successively trimming alternate 27 μm wide tracks as discussed in Chapter 7.

A flying erase head is mounted at 90° to the video heads and is 68 μm thick. This thickness allows it to erase two tracks simultaneously and provides the 8 mm format with a high grade of editing, without the colour flutter on assemble edits of previous formats and models.

Recording frequency spectrum

By using new types of video heads with extended performance figures both chrominance and luminance recording frequencies have been increased (see Figure 3.11h). The luminance sync tip frequency is 4.2 MHz and peak white is 5.4 MHz, with the upper side band extending well over 6 MHz giving improved luminance frequency response. Crosstalk between tracks is reduced by a $\frac{1}{2}f_h$ shift in the Ch.1 FM carrier and so the crosstalk cancellation system is similar to that used in Betamax systems, described in Chapter 3. The lower luminance sideband is filtered to just less than 2 MHz to avoid beats with the 1.5 MHz audio FM carrier. There is only a mono audio FM track; stereo is accommodated by the PCM audio system and is not used in the standard videomovie camera. The centre 1.5 MHz has a maximum deviation of 100 kHz but in practice it is restricted to 60 kHz. A limitation has been the inability to post-dub audio, except by adding PCM tracks using an advanced mains chassis 8 mm VCR.

Chrominance is centred on a 732 kHz subconverted carrier. Chroma crosstalk cancellation is achieved by an advancing phase shift of +90°/line in the signal recorded by Ch.1 video head. Although the 8 mm system indicates that a single frequency is used it can be argued that 3.90624 kHz is added to the subconverted frequency as given in the colour crosstalk section of Chapter 5. Subconverted colour carrier for Ch.2 is determined by the factor of $(46+7/8)f_h$ = 732.421880 kHz from a VCO at 5.86 MHz. Ch.1 frequency is therefore equivalent to 773.632880 kHz and crosstalk cancellation during colour replay is basically the same principle as for VHS, only the figures for carrier frequencies are different. In VHS, Ch.1 colour under carrier signal is kept in constant phase, except for the PAL swing and Ch.2 is phase retarded by -90°/line: this has exactly the same effect as the 8 mm system described which shifts Ch.1 by +90°/line whilst recording Ch.2 colour constant. The explanation for 8 mm colour recording and replay is therefore the same as VHS but with a change in the values. (see page 78.)

Automatic tracking

Automatic tracking first came out in the V2000 format with its DTF tone frequencies; the 8 mm format has a similar system using four ATF frequencies.

	Precise frequencies	*Rounded off*
f1	101.024 kHz	101 kHz
f2	117.188 kHz	117 kHz
f3	162.760 kHz	163 kHz
f4	146.484 Khz	146 kHz

The different frequencies used for ATF are 16 kHz and 45 kHz, and the operation is the same as V2000 except there is no DTF frequency burst for record track spacing.

During record the ATF reference frequencies are recorded in the sequence f1, f2, f3, f4. During replay, the reference sequence is changed to f4, f3, f2, f1, in order to determine the correction range of video tracking by use of the capstan servo. When the recorded ATF frequencies for ATF are replayed, on the other hand, they come off the tape in the original sequence f1, f2, f3, f4. Both replayed and reference frequencies are mixed, as shown in Figure 1.18. Reference frequencies are generated by the control system and their frequencies and sequence are programmed for different modes of operation such as cue, review and reverse play. The mixed result is fed to 45 kHz and 16 kHz filters, the outputs of which are detected and provide inputs to balance an amplifier.

When the video head is 'on track' the operational amplifier is balanced as both inputs are at equal level. If the video head is advanced so the 16 kHz signal is at a higher level than the 45 kHz signal then the op amp output falls and, of course, the reverse if the video head is lagging. A sample and hold circuit reads the error signal and holds it for the capstan servo, a number of sample pulses are used at different times to detect phase lock to ensure the correct sequence of reference pilot frequencies as well as the phase error.

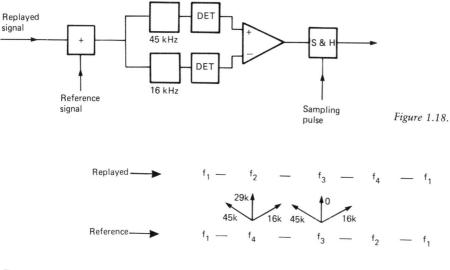

Figure 1.18.

Figure 1.19.

The matching sequence which determines correct tracking operation is shown in Figure 1.19. Note that f4 is referenced to f2 resulting in a centre frequency of 29 kHz which is ignored as it does not match any filter, only 45 kHz and 16 kHz difference beats are utilised.

An advantage of this type of ATF servo is very accurate editing, as any edits must maintain the correct replay sequence to avoid a disturbance at the cdit point which is achieved by using f1 as the edit switching reference signal. There is also the future possibility of editing over a PAL four field sequence, which is usually found on broadcast standard video recorders.

16

2

AZIMUTH TILT TECHNIQUE

The video tracks and guard bands can be seen in Figure 2.1a. One video head is shown properly on track and replaying maximum FM signal, the other is shown off the track with improper tracking. Although the second head is not scanning the full width of the FM magnetic track there will still be a reasonable quality replay for two reasons. First, there will be sufficient signal picked up from the magnetic track that is covered to enable the

As the VHS and Betamax systems were being developed in Japan for the American and Japanese markets, Philips were developing the N1700 format. The aim of these parallel developments was to produce formats with longer playing times. In order to achieve longer times, bearing in mind that the N1500 was limited to 1 hour, the tape would have to be slowed down and the magnetic track 'packing density' increased, that is to put more

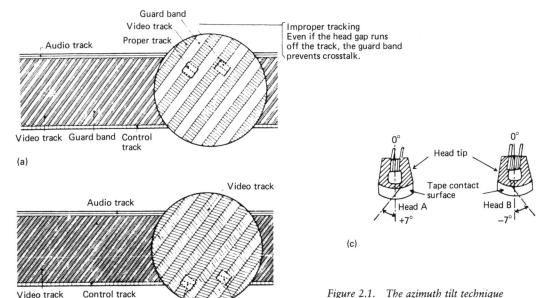

(a)

(b)

(c)

Figure 2.1. The azimuth tilt technique

replay amplifiers and limiters to operate satisfactorily and demodulate the signal. Secondly, the part of the video head that is off the magnetic track is scanning blank tape and *not* the adjacent field track.

If the guard band were not present and the video head wandered off its track onto the next one there would be a replay of both tracks together, i.e. a replay of two fields at the same time, and considerable crosstalk and patterning would occur. This is the reason for the guard band, at least originally.

information on the same area of tape. The most obvious development was to use the blank guard band between each track. Whilst the guard band served a useful purpose for protection against crosstalk it was in fact wasted tape space that could be better utilised to carry a signal. A technique was therefore developed which allowed the guard band to be eliminated, the azimuth tilt technique. To understand how it works you must appreciate that the FM carrier, onto which the luminance signal is modulated, is a high frequency signal.

Turning for a moment to audio tape recorders, for the best signal on replay the audio head must be vertical, that is at an angle of 90° to the tape. The adjustment to obtain maximum reproduction of the high frequencies, or the treble, from the tape is called the azimuth adjustment. If the audio head is not vertical, that is if it is tilted one way or the other then the efficiency of reproduction is impaired and the high frequencies are reduced, resulting in a loss of treble content.

The high frequencies are not reproduced by a head that is tilted at an angle different to the angle of tilt used in the recording mode. The technique applied to videorecorders is to cut the gaps in the video heads at an angle, one way for one head and the opposite way for the other head as illustrated in Figure 2.1c, shown for the Betamax system. Video head A records its video track with the azimuth of its air gap at +7° and video head B records its magnetic track with the azimuth slanted at −7°, resulting in a total difference of 14° between the two video tracks. In replay the result of this difference in azimuth is that each video head can only record and hence replay its own magnetic track of FM luminance carrier. For the high frequency FM luminance carrier each head can reproduce only its own track, the one that *it* recorded. This is valid for all current systems, VHS, Betamax, V2000 and N1700.

It can be seen from Figure 2.1b that the appearance of the magnetic tracks is like a herringbone pattern due to the azimuth tilt between the video heads; obviously this pattern is diagrammatic and cannot be seen in reality, except by special processes. The advantage now is that the tape can be slowed down and the guard bands filled with signal.

When a video head replays the track that it recorded the FM signal and the colour under carriers are reproduced normally. If that video head wanders off track slightly it will cover an adjacent track. The magnetic FM carrier of the adjacent track will have been recorded at a different azimuth angle by the other video head so that it will NOT be reproduced. However, some pick up will occur but it will be about 30 dB down, too low to create problems. However, there are noise reduction systems to reduce this carrier crosstalk still further (see the FM recording and replay advances in a later chapter). The colour undercarrier is a much lower frequency and is not affected by the azimuth slant so colour crosstalk will occur during replay and so colour crosstalk cancellation techniques have to be used to eliminate it (see Chapter 5). It later Betamax recorders there is an active luminance crosstalk comb filter to 'clean up' any luminance signal that is not sufficiently attenuated by the azimuth slant.

The JVC HR7650 and later versions of the HR7200 and HR7300 also have further luminance FM carrier crosstalk comb filtering techniques added, similar to those described on p. 41.

3

FREQUENCY MODULATION

The video signal is a wide bandwidth signal of varying amplitude, but for recording purposes it is converted to a signal of varying frequency on a high-frequency carrier. There are two main advantages in this technique.

The bandwidth of the signal is from near DC to 3 MHz which is greater than 12 octaves, and such a range cannot practically be handled by a magnetic recording head. By frequency modulating the signal onto a high band carrier and by using a high frequency video recording head the octave range can be greatly reduced.

The second advantage of utilising a frequency modulated carrier is that it is unaffected by amplitude variations, which occur due to head to tape contact problems in both record and replay, as the video heads are subject to bounce. Variations in the tape tension cause the tape to stretch and flex as it passes through the mechanics, also contributing to amplitude fluctuations of the FM carrier.

The bandwidth of the system is not flat over the recording frequency spectrum, high frequencies around the 4–5 MHz area will also cause a reduction in FM carrier level. An FM modulated carrier can be amplified and then limited to eliminate amplitude variations leaving a 'clean' FM carrier for demodulation.

Figure 3.1 illustrates the frequency modulation relationship. Sync tips correspond to the carrier free run frequency of 3.8 MHz, peak whites correspond to maximum modulation of 4.8 MHz, thus converting an amplitude modulated signal to a frequency modulated signal. The frequency spectrum of the video signal from DC to 3 MHz creates sidebands extending above and below the modulated carrier frequency of 3.8 MHz–4.8 MHz. These sidebands are limited to 1.3 MHz on the lower sideband and 5.75 MHz on the upper sideband.

The figures given are for the VHS system, with a total recording spectrum of 1.3 MHz to 5.75 MHz which is just over two octaves with the colour under carrier centred on 627 kHz (see Figure 3.5).

Frequency modulated recording

The three main advantages of FM recording are:

1. Permits the recording and replay of a wide band signal on a relatively narrow band video head. This means that a video signal with a bandwidth from 0 V (DC) to 3 MHz can be recorded using an FM carrier deviated from about 3.8 MHz to 4.8 MHz, these values depending on the format.
2. Replay amplitude variations arising from irregularities of head to tape contact can easily be removed by using an FM carrier limiter.
3. The AC bias signal, as required for audio tape recorders, is not needed as the FM carrier forms its own bias and the bias for the colour under carrier signal.

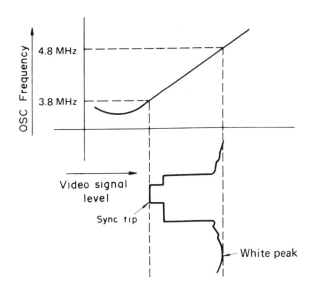

$$f = \frac{1}{2CR \ln \left(1 + \frac{E_o}{E_i}\right)}$$

Figure 3.1.

19

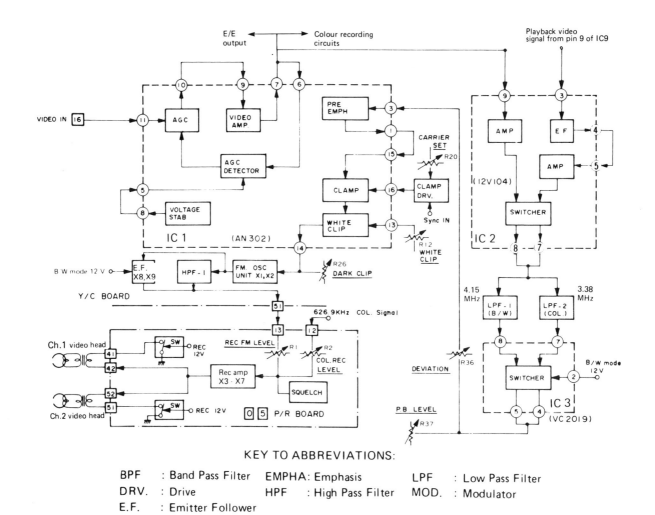

KEY TO ABBREVIATIONS:

BPF	: Band Pass Filter	EMPHA	: Emphasis	LPF	: Low Pass Filter
DRV.	: Drive	HPF	: High Pass Filter	MOD.	: Modulator
E.F.	: Emitter Follower				

Figure 3.2.   FM recording circuits of a typical VHS system

Prior to recording, the video signal has to undergo an amount of processing and the example we shall look at first is a typical VHS system (Figure 3.2).

A video signal from either the tuner/IF or the external input is fed to an AGC (automatic gain control) network which stabilises the mean or average level of the video signal. Correction is available against inputs from cameras, other recorders and variations in received signal level which are not at the standard 1 V p-p.

AGC output is split up into three paths as the signal contains full CVBS (Colour, Video, Blanking and Syncs). First it provides the E/E signal output to the UHF modulator and video output socket for monitoring purposes; note that when a recording is monitored all the viewer sees is the AGC output. Secondly the video signal is sent to

the colour processing circuits where chroma is filtered out for the colour recording signal. Lastly the signal passes via a record/replay switcher to a pair of filters, one for monochrome signal recording which cuts off at 4.15 MHz and the other for colour recording which cuts off at 3.38 MHz to filter out the 4.43 MHz colour signal which may cause beat patterning with the FM carrier. Either filter is selected by a switcher working off the colour killer switching signal, or manually, by a black and white selector switch. The filters are used in dual roles in both record and replay paths; their response curves are shown in Figure 3.3.

Out of the filter selecting switch, VC2019, in record mode the signal travels to a level control which determines the amplitude of the signal to the FM modulator circuit. In the modulator this

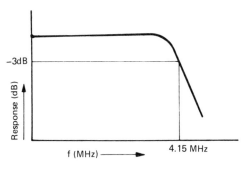 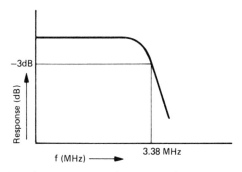

Figure 3.3. Filter response curves; left, LPF-1 (black and white mode); right, LPF-2 (colour mode)

control sets the peak white FM carrier frequency of 4.48 MHz. It is therefore called the FM deviation control. The signal passes next into an IC (AN302) where pre-emphasis is added which makes it rather spiky with overshoots at peak whites and sync tips (Figure 3.4).

After pre-emphasis the signal is clamped so that the lines sync pulse tips are held at a clamped DC level into the FM modulator corresponding to 3.8 MHz. The precise frequency can be set by the 'set carrier' control to adjust the clamped level.

Video signal amplitude thus extends in the positive direction up from this level, stabilising the signal to the FM oscillator against DC drift. Before the video signal is allowed to get to the modulator the overshoots created by the pre-emphasis stage have to be removed to protect the FM oscillator from overmodulation. Figure 3.4 also illustrates the problem associated with overmodulation. Peak white overshoots drive the modulator to a frequency much higher than that which the record/replay circuits and the video heads can handle. The amplitude of this high frequency is too low for the limiter to amplify, that is it is below the clipping level. In replay, then, there will be part of the FM carrier missing and the resultant output will be 'black'. As we shall see, in replay mode there is a circuit to alleviate this problem under normal working conditions. In abnormal conditions, such as incorrect video record signal levels or worn video heads, the correction circuits cannot cope. The visual effect in replay is black speckling on white edges, most moticeable on mid-afternoon transmissions of horse racing when white titles of high amplitude appear. A high pass filter takes the modulated signal from the modulator to the record amplifier where it is joined by the chroma colour under carrier record signal.

Figure 3.5 shows the spectrum of the composite video signal converted to the spectrum of the FM and colour under carriers for recording, the colour signal being centred around 627 kHz and the FM carrier between 1.3 MHz and 5.75 MHz.

The record amplifier and its response curves are shown in Figure 3.6. It can be seen from the graph of playback output versus record current that there is an optimum current that passes through the record/playback curves at different frequencies. This current is set by the record FM carrier level control, the colour record current is also optimised by the colour record level control.

Figure 3.4.

21

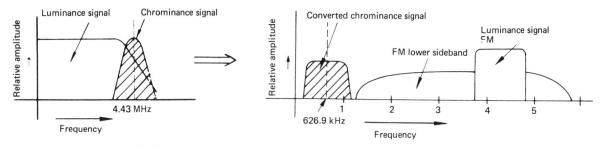

Figure 3.5. Frequency response curves: left, PAL spectrum; right, converted FM spectrum

The most difficult setting up procedure in the record amplifier circuits is that of the FM carrier deviation and carrier levels; however, with a little thought and the correct equipment it can be done. Equipment required is an RF generator to cover 3–5 MHz, a video pattern generator with a white pattern, grey scale or a chess board pattern could also be used, two 1 k resistors and a dual beam oscilloscope. As this may need to be carried out on any format video recorder the procedure is generalised.

Locate a point within the video recorder which corresponds to the output of the FM modulator through a buffer stage, this is usually the output of the luminance record panel or the input side of the video head drive amplifier before the colour under carrier is added.

Connect this point via a 1 k resistor to an RF generator which is also feeding through a 1 k resistor. Then connect an oscilloscope to this junction by one of its inputs (Figure 3.7a). The other input to the oscilloscope is connected to the video output socket.

Connect a pattern generator to the video recorder's line input and put the machine into the 'record' mode.

Synchronise the oscilloscope to the video signal at line rate and compare the two beams; set the RF generator frequency and level to display zero beat signal on the edge of the FM carrier that corresponds to the line sync tip (Figure 3.7b). Once reasonable conditions for the display have been established set the RF generator to the sync tip frequency, which would be 3.8 MHz for a VHS

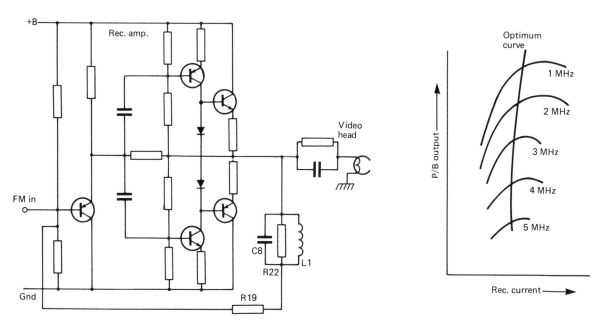

Figure 3.6. The record amplifier and its response curves

recorder, then adjust the clamp level or 'set carrier' control for minimum beat in the area of the sync tip/FM carrier.

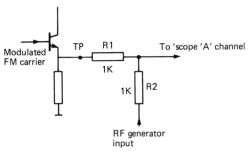

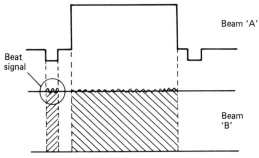

Figure 3.7.

Re-adjust the RF generator for peak white frequency, 4.8 MHz in the case of VHS, and examine the edge of the FM carrier in the area corresponding to the peak white signal; adjust the video AGC level for zero beat in this area.

Check that the video level in from the pattern generator is 100% peak white or that it corresponds to peak white off air, if not re-adjust the level pot accordingly.

The frequency modulator

The frequency modulator is shown in Figure 3.8. A tuned transformer forms the output stage with a secondary winding to feed the FM carrier to the record drive amplifiers. Tuning is done by capacitor C_t and as it can influence the frequency of oscillation it will operate in conjunction with the 'Set carrier' DC clamp level potentiometer. C_t should not normally be adjusted as twiddling can severely alter the FM modulator range and cause attenuation at the higher frequencies. C_t tunes the transformer for maximum bandwidth in the range of 3 to 5 MHz and is set for maximum output in this range. Without a sweep oscillator and the correct instructions C_t can only be set by an engineer with considerable experience.

The main adjustment in the FM oscillator is the potentiometer R_s which ensures that there is symmetrical drive to the pair of transistors in the oscillator. It is set by viewing the output on an oscilloscope with no modulation so that each cycle of the oscillations can be seen clearly and then adjusting for the most symmetrical waveform, so that the positive half of the cycle is a mirror image of the negative half. Note that you will not see a pure sine wave; it will have small imperfections.

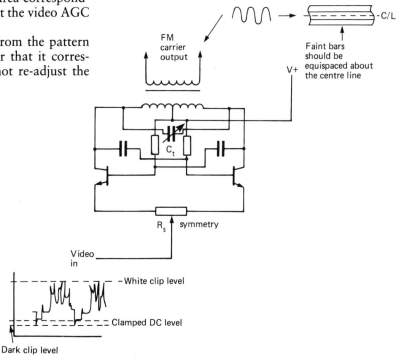

Figure 3.8. An FM modulator

23

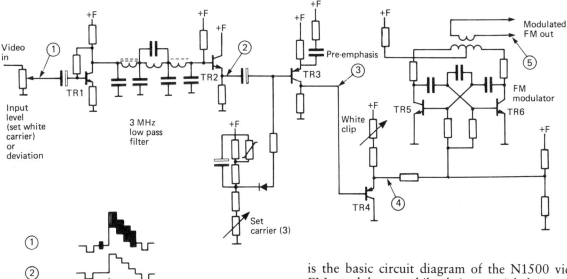

Figure 3.9. N1500 FM modulator

On some machines, the waveform is best displayed as a carrier band using a lower oscilloscope sweep rate; in this way faint horizontal bands can be seen above and below the centre line of the band. These bands are caused by the imperfections in the waveform, and R_s is adjusted so that the faint bands are symmetrical about the centre of the FM carrier, above and below its centre line. The video signal applied to the modulator is also shown, the clamped DC level, and the peak white clip and dark clip levels which trim off the overshoot spikes, created by pre-emphasis, top and bottom of the waveform can also be seen.

So much for FM recording theory; let us now consider a practical approach on circuits from the Philips N1500 using discrete components. Some simplification has been made in order that the techniques are clearly seen. Modern videorecorders have the signal processing carried out inside integrated circuits, so much so that the circuit diagram starts to look more like a block diagram and the real signal processing is hidden. Figure 3.9

is the basic circuit diagram of the N1500 video FM modulator; whilst being straightforward it does contain the necessary features for FM recording.

Video comes in from the tuner/IF circuits via a level control. Note that there is no automatic gain circuit. The N1500 did not have any auxiliary or external inputs so it relied on the IF AGC to produce a fixed level video signal to the recording amplifiers. The input level control determines the amplitude of the signal, and at this point it is a full colour bar signal but it passes through 3 MHz low pass filter which limits the luminance bandwidth and removes the colour content. The filter section is buffered by an emitter follower, TR2, which provides low impedance drive to TR3. The base of TR3 is clamped to a set DC voltage, adjusted by the 'set carrier' control and temperature compensated against drifting. From TR3 to the modulator the signal path is DC coupled and so the voltage on the base of TR3 sets the free running frequency of the FM modulator multivibrator. If any drift in the TR3 components were to take place then the frequency of the modulator would alter; this in turn would change the range of frequencies that form the modulated FM carrier, which in this case are 3 MHz (sync tip) to 4.4 MHz (peak white).

Pre-emphasis is applied by the use of the decoupling components in TR3 emitter, this creates the 'spiky' signal with lots of overshoots in it. The peak white overshoot is removed by TR4, whose emitter is held at a fixed DC voltage which is adjusted by the 'white clip' potentiometer. The emitter of TR4 cannot go any more positive than this set voltage; peak white overshoots applied to the TR4 base which are more positive than the fixed emitter voltage will cause TR4 to cut off and so limit the value of peak white. Note that there is no such similar circuit for 'dark clip' to limit the sync tip negative overshoots. The modulator range can tolerate these so a dark clip appears on mainly Japanese recorders.

The varying level on TR4 emitter formed by the video signal modifies the frequency of oscillation of TR5 and TR6, and the output of the oscillator is taken from a secondary winding on the tuned transformer. This is the modulated FM carrier which now has to be mixed with the colour under carrier and driven to the video heads by a special recording amplifier.

The FM recording amplifier has to add the luminance FM carrier to the colour under carrier at their correct optimum respective levels and provide symmetrical low impedance current drive to the video heads as well as give some HF lift to overcome head to tape bandwidth reductions. The circuitry is shown in Figure 3.10. The FM signal input will contain some amplitude variations due to the impedance of the coupling transformer and the oscillator causing level variations with frequency changes, also there will be a length of screened cable from the modulator to the record drive amplifier which will be susceptible to pick up. Diodes D1 and D2 clip the FM signal top and bottom thus removing such variations.

Prior to mixing both the FM and the colour under carriers undergo HF lift and resistive mixing is used. A class B amplifier provides recording current drive which is measured at TP1, across a very low value resistor (typically 1 ohm). The measurement is in mV, and the colour is adjusted first by shorting the luminance to chassis across test points at the level control to remove the FM

carrier. This to enable the colour signal which is very small to be seen clearly at around 2 mV. The FM record current is then set at around 25 mV. Great care has to be taken to terminate the coax leads to the meter at 75 ohms, and a wide band millivoltmeter is specified. The author uses the oscilloscope and sets the levels at 5 mV for colour and 70 mV for luminance (peak to peak values).

The FM recording drive on some machines could be optimised to a particular tape, especially as newer high energy tapes are appearing on the scene; the method is as follows.

The recorder is set up in record mode with a microphone connected to the audio input and colour bars to the video input. An oscilloscope is connected to the FM drive amplifier output; the level will be found to be around the 2–3 V mark. The FM drive is then turned down to a suitable minimum, for example ¼ V, it is then advanced ¼ V at a time whilst announcing the value into the microphone up to a maximum of 4 V. Having completed this, the tape is then replayed with the oscilloscope connected to the replay pre-amp output. The replayed FM carrier will then increase with each announced record level and after reaching a peak value it will fall again as the recording level is further increased. All you do then is set the recording level to that value which produced the maximum replay level. Some recorders will flatten off around certain values, some VHS recorders will not alter the replay level value between 2½ V and 3½ V recording level. Their record drive levels are given as 3 V p-p which is usually close enough.

The colour under carrier record level is best set out as follows. Replay the manufacturers' test tape at the colour bar section and measure the replay

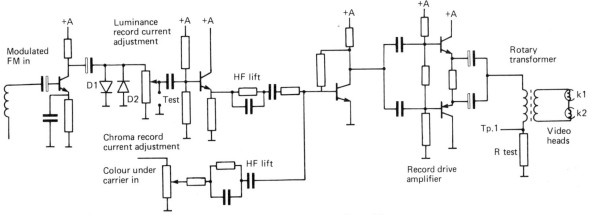

Figure 3.10. N1500 FM record amplifier

25

colour level on the output of the replay pre-amp – do not measure it on the colour PCB as it may have gone through an ACG level control amplifier. Having noted this replay level, record some colour bars, preferably not on the test tape – don't forget to put a new one in the machine! It is then a matter of recording and replaying colour bars, each time adjusting the colour record level until such time as the replayed level meets the value measured off the test tape.

A comparison of FM carrier frequency spectra

Within the FM carrier spectrum two main parameters are set, one is the bandwidth of the video signal and hence the resolution of the replayed picture, the other is the signal to noise ratio of the video signal. Bandwidth is determined by the range of the FM lower sideband between the lower limit around 1 MHz to the centre frequency of the modulation band. Signal to noise is determined by the bandwidth of the modulation frequency and the coercivity of the video tape.

In the specifications laid down for the early Philips, VHS, Betamax and V2000 the video tape available was of a low coercivity, some 400–500 oersteds in general. After development of higher coercive tape and when Betamax was under threat from VHS, Sony announced a Super Beta system to provide high quality pictures and compete with VHS. VHS was nearly ten years old when JVC announced a Super VHS or S-VHS system with near broadcast bandwidth.

TV lines

Video signal bandwidth can be looked at in two ways. The first is equivalent to the audio specification of -3 dB points with respect to a sine wave. This is never quoted, however, as it would appear very low indeed and anyway, enhancement techniques are typically used which can produce a picture far better than a consideration of only the signal's bandwidth would convey.

The other way is to consider bandwidth in terms of picture resolution. Resolution is quoted in lines which can be resolved by an observer per picture height. Although the topic of picture resolution is highly subjective, 625-line TV tubes are generally

considered to have a resolution of about 400 vertical lines.

When referring to VCRs, because resolution *is* so subjective then an approximation (1 MHz to 80 lines) relating bandwidth to the number of lines which can be resolved is all that is required. VHS, with its bandwidth of about 3.2 MHz, for example, allows a resolution of about 250 lines.

N1500

In the first Philips system (Figure 3.11a) the FM carrier was modulated between 3 MHz at the sync tip and 4.4 MHz at peak white, peak white clip level approximated to about 4.8 MHz. Colour under carrier was set at 562.5 kHz and it had to be a multiple of line frequency. Line frequency f_h, 15.625 kHz multiplied by 36 is 562.50 kHz.

N1700

This is the second Philips system (Figure 3.11b). Introduced into this videorecorder was a dark clip or black clip, which limited the lower level of the FM carrier to 2.9 MHz whilst the peak white clip was limited to 4.8 MHz as in the N1500. Colour under carrier was the same as the N1500 being f_h × 36.

VHS

The VHS system (Figure 3.11c) has a colour under carrier about $40f_h$ which includes a slight frequency shift, $40f_h$ is 625 kHz, the shift being 1.952 kHz results in a carrier of 626.952 kHz. The FM modulation is between 3.8 MHz and 4.8 MHz and the modulator incorporates both dark and peak white clip adjustments. It can be proved that the colour under carrier is in fact two frequencies, for details refer to Chapter 5.

Betamax

The diagram for Betamax (Figure 3.11d) looks more complex than it really is, the colour under carrier is shown as two frequencies: 685.5 kHz and 689.5 kHz. Each carrier is allocated to a video head, Ch.1 being 685 kHz and Ch.2 being 689 kHz.

The FM carrier is between 3.8 MHz and 5.2 MHz, this probably indicates the greater bandwidth capabilities of Betamax. Again both dark and peak white clip levels are used.

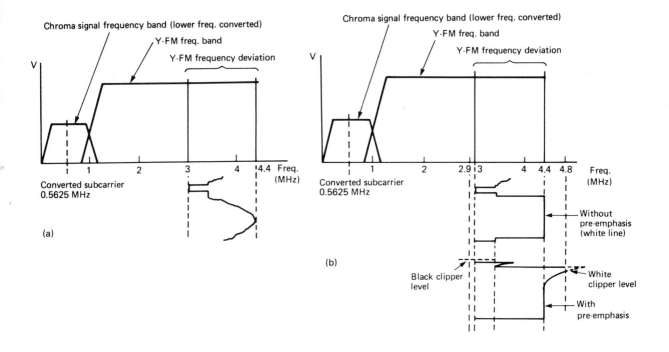

(a)

Chroma signal frequency band (lower freq. converted)
Y-FM freq. band
Y-FM frequency deviation

V

1   2   3   4   4.4 Freq. (MHz)

Converted subcarrier
0.5625 MHz

(b)

Chroma signal frequency band (lower freq. converted)
Y-FM freq. band
Y-FM frequency deviation

V

1   2   2.9 3   4   4.4   4.8   Freq. (MHz)

Converted subcarrier
0.5625 MHz

Without pre-emphasis (white line)

Black clipper level

White clipper level

With pre-emphasis

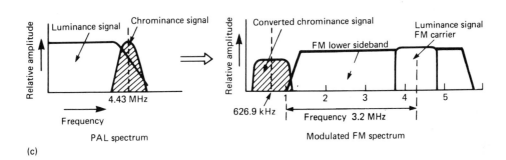

Luminance signal
Chrominance signal

Relative amplitude

4.43 MHz

Frequency

PAL spectrum

Converted chrominance signal
FM lower sideband
Luminance signal FM carrier

Relative amplitude

626.9 kHz   1   2   3   4   5

Frequency 3.2 MHz

Modulated FM spectrum

(c)

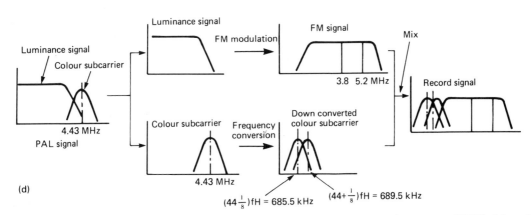

Luminance signal

FM modulation

FM signal

Mix

Luminance signal
Colour subcarrier

4.43 MHz
PAL signal

Colour subcarrier

4.43 MHz

Frequency conversion

Down converted colour subcarrier

$(44\frac{1}{8})fH = 685.5$ kHz

$(44+\frac{1}{8})fH = 689.5$ kHz

3.8   5.2 MHz

Record signal

(d)

Figure 3.11.   Comparison of FM carrier frequency spectra. a, Philips N1500. b, N1700. c, VHS. d, Betamax. e, V2000. f, Super Beta. g, Super VHS. h, 8 mm video

27

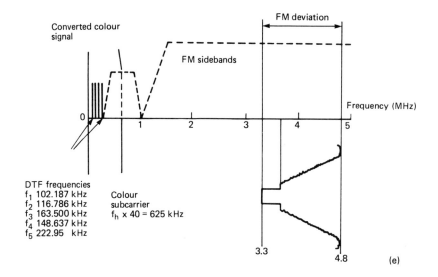

Converted colour signal

FM sidebands

FM deviation

Frequency (MHz)

0   1   2   3   4   5

DTF frequencies
f_1 102.187 kHz
f_2 116.786 kHz
f_3 163.500 kHz
f_4 148.637 kHz
f_5 222.95 kHz

Colour subcarrier
$f_h \times 40 = 625$ kHz

3.3   4.8

(e)

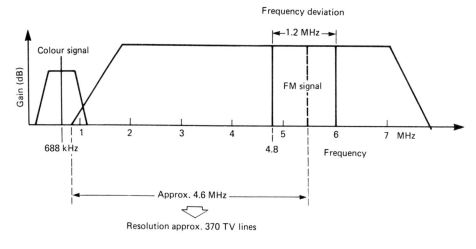

Frequency deviation

1.2 MHz

Colour signal

Gain (dB)

FM signal

688 kHz   1   2   3   4   5   6   7   MHz

4.8   Frequency

Approx. 4.6 MHz

Resolution approx. 370 TV lines

(f)

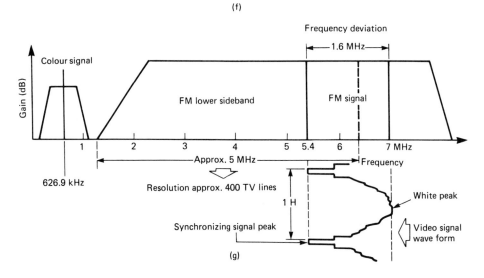

Frequency deviation

1.6 MHz

Colour signal

Gain (dB)

FM lower sideband

FM signal

1   2   3   4   5   5.4   6   7 MHz

626.9 kHz

Approx. 5 MHz

Frequency

Resolution approx. 400 TV lines

1 H

Synchronizing signal peak

White peak

Video signal wave form

(g)

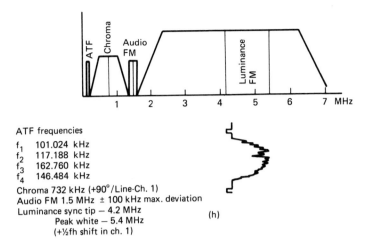

ATF frequencies

f_1 101.024 kHz
f_2 117.188 kHz
f_3 162.760 kHz
f_4 146.484 kHz

Chroma 732 kHz (+90°/Line-Ch. 1)
Audio FM 1.5 MHz ± 100 kHz max. deviation
Luminance sync tip – 4.2 MHz
          Peak white – 5.4 MHz
          (+½fh shift in ch. 1)

(h)

V2000

In this system (Figure 3.11e) the colour under carrier is chosen as $40 f_h$ (625 kHz) and the FM carrier modulated between 3.3 MHz and 4.8 MHz. Note that the DTF frequencies are slotted in at the lower end of the spectrum so as not to create intermodulation components which would appear as patterning.

Super Beta

The spectrum for Super Beta is shown in Figure 3.11f, and while the down-converted colour frequency is the same as for standard Betamax the FM modulation frequency has been moved up the band. FM deviation has been increased to 1.2 MHz, from 4.8 MHz sync tip to 6 MHz peak white to improve the signal to noise ratio. The lower side band has been extended to 4.6 MHz to provide a resolution of 370 lines. A tape recorded on Super Beta cannot be replayed on a standard Betamax video recorder.

Super VHS

As in the Super Beta system, Super VHS has retained the same centre colour frequency, while the FM bandwidth has been increased. More important there is a gap of about 300 kHz between the colour upper sideband and the luminance lower sideband to reduce luma/chroma crosstalk. To improve signal to noise ratio the FM deviation is increased to 1.6 MHz in a band of 5.4 MHz sync tip to 7 MHz peak white which allows for a lower sideband increase to 5 MHz i.e. 400 lines resolution. See also VHS HQ section at the end of this chapter.

8 mm format

The 8 mm video system spectrum is shown in Figure 3.11h. FM deviation is 1.2 MHz between 4.2 MHz and 5.4 MHz with a lower side band of about 3.3 MHz i.e. of about 270 lines resolution. The resolution restriction is limited by the inclusion of an FM audio band which is part of the recording spectrum and not recorded by a separate audio head. The colour system is similar to VHS but at a different frequency and the automatic tracking system is similar to V2000 but is used in a more complex way.

FM replay and demodulation

In replay, each video head scans the tape in turn, replaying the magnetic tracks and reproducing the FM carrier signal one field at a time. It is an important note, at this point, to remember that the videotape wrap around the video head drum is in excess of 180°. In replay this will provide for an overlap of replayed signal from each head, to avoid any discontinuity of the replayed FM carrier. More of this point later.

Firstly, a look at the overall block diagram of a typical replay system taken from a standard VHS recorder, model HR3300, in Figure 3.12.

The replayed FM carrier is amplified in a tuned pre-amplifier, the response being tailored to colour or monochrome. This switching of colour and monochrome was discontinued in later machines. A drop out compensater is used in this part of the system. A drop out is a white spot or even a

29

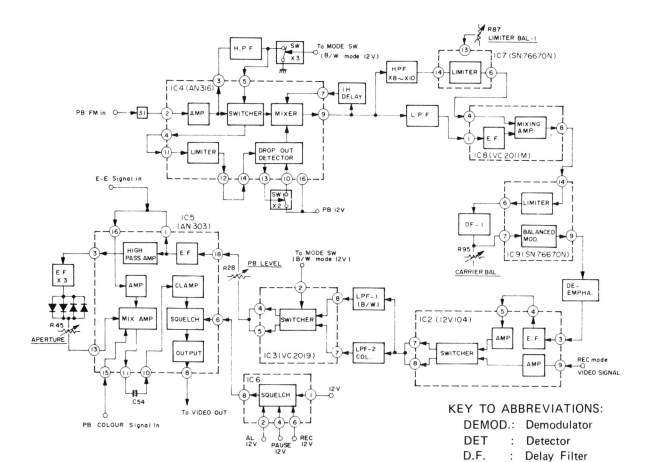

KEY TO ABBREVIATIONS:
DEMOD.: Demodulator
DET   : Detector
D.F.   : Delay Filter
E.F.   : Emitter Follower
HPF   : High Pass Filter
LPF   : Low Pass Filter

Figure 3.12.   Block diagram of HR3300 (VHS) replay system

Key to abbreviations
COL: Colour   LPF: Low pass filter
EQ : Equaliser   SW : Electronic switch

Figure 3.13.   Input stages of replay system

30

horizontal line caused by magnetic particles missing from the tape itself. A one line delay line is used, in the event of a drop out occurring then the missing FM carrier is 'filled in' by carrier stored in the delay line from a prior replayed line. After the drop out compensator considerable limiting is employed to eliminate any amplitude variations in the signal resulting from head to tape bounce and as the VCR ages, video head wear. A balanced modulator (IC9) is employed to demodulate the FM carrier, the resultant signal being a video signal full of carrier, which is then filtered out by a low pass filter, which doubles up as de-emphasis.

The resultant video signal is switched and via a level control is fed to an aperture corrector – a crispener circuit to enhance the black to white, and white to black transients. It is sometimes called edge enhancement but too much enhancement will create overshoots and ghosting, or reflections.

After aperture correction the replayed chroma signal is added, the composite signal is then buffered and fed to the modulator via a squelch circuit, which blanks the video signal for a few seconds after the play button has been selected. This allows the servo circuits to settle before the video is allowed to the output.

The pre-amplifier and video head switching

The pre-amplifier is split into two input stages (Figure 3.13), usually a low noise field effect transistor, or low noise IC, one for each video head. Although the video heads are coupled to the pre-amplifiers via a single rotary transformer, they are not interactive. This is because the rotary transformer is split into two separate sections each having two close coupled windings, an upper winding connected to the video head, which is rotating, and a lower winding connected to the pre-amplifier, which is static This is in effect two rotary transformers in one, one rotary coupling for each video head. Across the secondary of the rotary transformer is the Q network, to peak the response of the video head and rotary transformer to a frequency of 5.0 MHz (see Figure 3.14).

It is necessary to align each of the Q networks each time a head assembly is replaced; they must be balanced for frequency and amplitude. If the Q networks for each video head are not balanced a

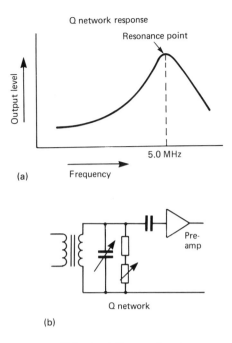

Figure 3.14. *Video head Q network and response curve*

selective flashing effect can be seen on the replayed picture, which under normal circumstances cannot be readily defined or explained. However, replay of frequency graticules as can be obtained from the more expensive pattern generators which reproduce bands of vertical lines from 0.5 MHz to 3.0 MHz, will define the flashing. The effect, on replay of such sets of frequency gratings, of unbalance in the Q networks will be that the 25 Hz flashing will be confined to certain frequencies only. This explains why flashing on a replay of moving pictures will result in the flashing effect moving around the screen, whenever the frequencies at which the imbalance occurs are reproduced.

It is important not to meddle with the Q of each pre-amp unless you have a manufacturer's test tape which contains a sweep frequency section. A frequency sweep is the best way to peak 5 MHz or 4.5 MHz with the capacitor and then balance the amplitude with the pre-set damping resistors. It is normal to have at least one of the two variable damping resistors set at maximum resistance.

The outputs of the pre-amps are connected across two variable resistors, one is used to tap off the replayed colour under carrier (R49) and balance its level to be equal from each head. Setting up must be done replaying a replayed standard test tape again.

The other variable (R48) taps off replayed FM carrier to a second variable (R55) which adjusts the overall level of the mixed carrier. The replayed colour under signal is fed through a low pass filter to remove FM carrier, level adjusted and then buffered out to the colour replay processing circuits. Replayed FM carrier is passed through an equalising stage to peak up the HF upper sideband, this is additional to the Q tuning which aids S/N ratio.

Each video head replays more FM carrier than is actually needed to provide an overlap period. In this overlap period the signal information is duplicated so it is only necessary to switch one head off and the other head on to maintain continuity of the replay signal. The demodulated video signal will not then have any gaps in it where parts of the picture would be missing. There is only the head crossover point where the transition from one head to the other occurs. The crossover point is positioned 6–8 lines prior to field syncs, depending on the system. On the replayed picture the crossover point is at the bottom of the screen, out of sight in the overscan.

Figure 3.15 illustrates the Ch.1 and Ch.2 video head replayed signals. Switching is provided by the flip-flop signal, which is derived from the rotating drum, in the servo circuits.

Transistors Q1 and Q2 are switched 'on' 180° out of phase. When the flip-flop is 'low' then Q2 is on and Ch.2 video head is clamped to common supply. Ch.1 video head is replaying out via R1 (colour under) and R2 (FM carrier). At the changeover point both heads are reproducing FM carrier and colour under carrier so there is no loss of signal and there is a small amount of tolerance of the switching point, ±3–4 lines.

The accuracy required for the switching point is why the switching edges of the flip-flop have to be adjusted after a head change, again using a manufacturer's test tape as a reference.

A comparison of replay head switching and amplifiers is shown in Figure 3.16. The original Philips N1500 had only one rotary transformer winding because the video heads were in series. In consequence only a single pre-amplifier is used with the Q network fixed. There was no active video head switching and a sufficiently good signal to noise ratio was achieved in spite of the alternate head producing noise in addition to the scanning head.

VHS and Betamax are very similar as they were developed in the same era. Both have dual rotary transformer windings and Q networks and pre-amps switched from a flip-flop, or RF switching signal as it is sometimes called. The V2000 systems produce the most complex switching; the example is from Grundig 2 × 4 Plus and illustrate an important point mentioned in Chapter 5. First, in the record mode, S2a and S2b are both closed to ground. Record FM carrier is driven to the centre of the two windings so that both video heads are driven continuously in parallel. In replay S1 is closed, S2a and S2b are open and RF switching signal is applied in antiphase to switches S3a and S3b. When head Ch.1 is replaying S1 and S3a are closed and the signal is transferred to the single replay amplifier via the upper loop. When head Ch.2 is replaying S1 and S3b switches are closed and the signal is transferred via the lower loop.

From this information it is possible to deduce from the fact that both heads are recorded in parallel that during replay Ch.2 head is inverted with respect to Ch.1 head. Whilst this 180° phase

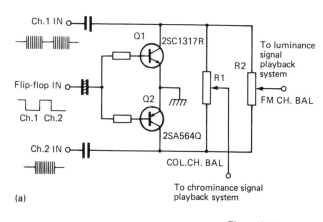

(a)

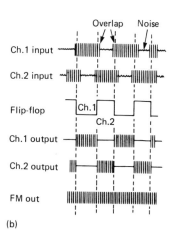

(b)

Figure 3.15.

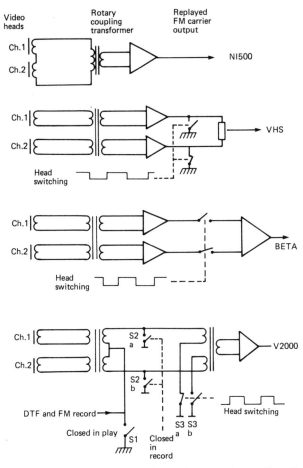

Figure 3.16. *Comparison of replay head switching and amplifiers*

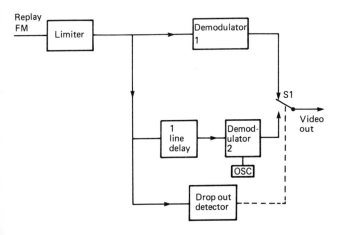

Figure 3.17. *Conventional drop out compensator*

inversion does not affect the FM carrier or its demodulation it does have a bearing on the colour under carrier. It causes the replayed colour under carrier to be inverted when replayed by Ch.2 video head. A correction switching signal at 25 Hz is applied to the replay colour processing circuitry to compensate for this inversion (see Chapter 5).

The drop out compensator

A drop out compensator is used to reduce random white spots which would be visible on the TV screen during replay. The engineer has only to inhibit the drop out compensator to see the quantity of drop out spots, which will increase with the age of the tape or its environmental storage conditions. Drop outs occur due to the tape shedding oxide particles and even new tape contains a certain amount of drop out. The drop out compensator, DOC as it is usually called, relies on the fact that the drop outs are random in nature and that they will not occur at the same position visually along subsequent lines, which would produce an effect like vertical white lines.

Figure 3.17 is a standard type of drop out compensation system used on U-matics and the Philips N1500 machines. The direct path is through a limiter and demodulator 1, via switch S1 and out as a video signal. A parallel path is through a 64 µs, one-line glass delay line and demodulator 2; the FM carrier is also fed to a drop out detector circuit, which monitors the FM carrier. When a drop out occurs an output switching signal is developed which is of very short duration lasting not much longer than the drop out and S1 changes over for the duration of the drop out. When switch S1 changes over the output from demodulator 2 is fed out as a video signal and 'fills in' for the drop out duration.

This small piece of video signal which is switched in, in place of the drop out, was replayed one line earlier and so it is used twice and prevents an obvious white spot, or black spot. Demodulators 1 and 2 are either matched or have separate gain and black level adjustments. In the extremes of drop out, when it is excessive due to high oxide shedding, it is possible for the delayed signal, which is switched to the output, to also contain drop outs. If this is the case then demodulator 2 is designed to oscillate at about 3½ MHz, to produce a mid-grey video level which is less obtrusive than pure white or black spots.

Cyclic drop out compensator

The example of the cyclic drop out compensator in Figure 3.18 is taken from a Grundig V2000, 2 × 4 Plus videorecorder although a similar system was used in the Philips N1700 series.

The direct signal path for the FM carrier is into IC_1, pin 5, through a limiter and demodulator to pin 16. Between pins 16 and 7 is a low pass filter to remove residual carrier from the video signal. Switch S1 connects the demodulated video signal to an output amplifier, having as part of its negative feedback path a de-emphasis network,

and the video signal is then sent to a variable enhancement circuit.

The parallel path is into IC_2 through pin 7 and switch S2 to the $64\mu s$ delay line, the delayed output passing back into IC_2 at pin 5 and into a limiting amplifier. The limiter has two outputs for the FM carrier, one to a demodulator and the other to switch S2 via a buffer amplifier. The demodulated and delayed video signal passes through another low pass filter to switch S1. Both switches S1 and S2 change over, driven by a drop out detector for the period of the drop out, and the detector is a schmitt trigger circuit, again within an IC (not shown).

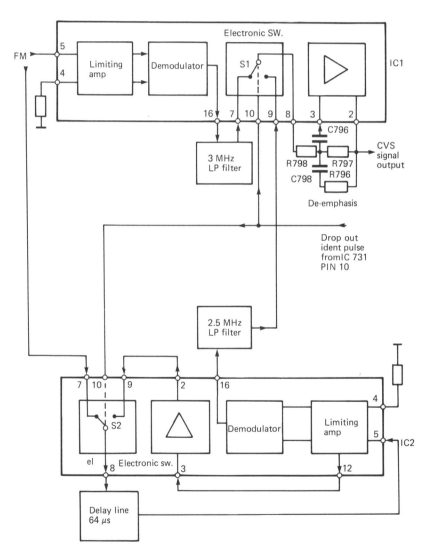

Figure 3.18. Cyclic drop out compensator

During a drop out, S1 switches between direct and delayed luminance video, the de-emphasis and output amplifier is common to direct and delayed video. This is similar to the basic drop out compensator in so much as the drop out is filled with delayed video. Also during a drop out S2 changes over and the output of the delay line is re-fed back to its input, thus preventing the delay line from being fed with a drop out by re-circulating its output.

The delay line is then always full of FM carrier without any drop out. However in the limits of very poor tape both switches may stay over, the delay line recirculates the FM carrier continuusly and the same video line is continuously demodulated and repeated at the video output. This is confirmed by the fact that a blank tape playing in a 2 × 4 series machine is displayed as vertical grey and white stripes from the last video line being

recirculated. The same effect can be seen as mentioned in the Philips N1700 and some later versions of the N1502, indicating the usage of a cyclic drop out compensator.

Double limiter

If the level of the FM carrier is too low for a limiter to fully amplify and limit the output, then a severe loss of carrier results, the lost signal when 'demodulated' being seen as a black on the television screen. This effect is seen when the FM record/replay frequency response is insufficient or the peak white clip level is set too high. A peak white will overmodulate the carrier and very high frequencies will be produced by the modulator.

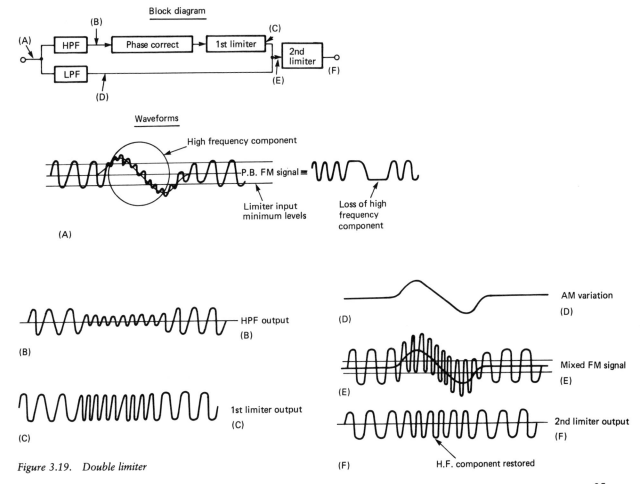

Figure 3.19. Double limiter

35

Bandwidth limitations will reduce the amplitude of the carrier and in replay the carrier will be too low for the limiter. As it is the peak white signals which create the highest frequency FM carriers, the visual effect on the screen during replay is black speckles after whites, most clearly seen on Philips N1500 when the video heads are wearing low. The speckling is less noticeable on VHS videorecorders due to a double limiter technique as shown in Figure 3.19 (A–F).

The waveform into the double limiter is shown (A): the high frequency component of the FM carrier, corresponding to the peak whites, is mixed onto a lower frequency component, forming a complex signal. Without double limiting only the low frequency component will be present at the limiter output as shown next to it, the high frequency component is lost and black is the result.

The double limiter technique is seen in the small block diagram; the high frequency component is split off through a high pass filter and is shown in waveform (B). It is phase corrected for shifts introduced by the filter and then limited to give waveform (C) which is then mixed with the low frequency component (D) to produce the complex waveform (E). The complex signal is then further limited in a second limiter and the output is as shown in (F); the high frequency component has been recovered and this means that (F) looks like (A) should have been without the loss of the HF component.

The VHS system can cope with a wide range of high frequency carrier levels and it is not subject to black after white speckling, except in the extreme cases of video head wear, that V2000 and Beta systems will display on white lettering.

The best all round test for this is recording mid-afternoon racing where the white titling is often not kept within the broadcasting maximum limits.

Demodulation

The FM carrier in record mode is frequency modulated between 3.8 MHz and 4.8 MHz, for VHS; from this carrier the modulation frequencies up to 2 MHz have to be recovered by demodulation and filtering. In order to simplify the design of the filters the FM carrier is effectively doubled in frequency to produce 7.6 MHz to 9.6 MHz. The filters needed to remove residual carrier are designed as low pass filters around 3 MHz to pass the demodulated video and reject the carrier; for a 3 MHz bandwidth signal a better rejection is obtained from a higher frequency 8 MHz carrier than one at 4 MHz.

A VHS demodulator is shown in Figure 3.20, and the basis of the demodulator is a fixed delay line, DF-1, with a very short delay time T_d. The FM carrier enters the demodulator IC on pin 14 where it is limited; the carrier is then split into two paths to an additive mixer, one path direct and the other via the delay line DF-1. It can be seen that this delayed signal, when added back to the direct signal, doubles the frequency. In Figure 3.20, (A) shows the input carrier waveforms at three sample frequencies; $T_o = 1/FC$, which is the centre frequency, $T_1 = 1/FC + \Delta F$ which the highest frequency and $T_2 = 1/FC - \Delta F$ which is the lowest frequency. If we deal with the centre frequency $T_o = 1/FC$ then the others can easily be followed.

(A) is the input direct to the additive mixer and (B) is the output of the delay line and the second input to the mixer, so (B) is (A) delayed by the short time T_d. In the mixer the two inputs are added and the castellation waveform (C) results; this derived waveform is then passed to a full wave rectifier to give a higher frequency output (d) which is double the input frequency (A) but does not have a symmetrical mark/space ratio. An alteration of the symmetry of the rectified output occurs. When passed through de-emphasis and filters it is smoothed to a DC level, E_o. As the carrier is frequency modulated the output E_o will vary in level according to the modulated frequency shift and will be the recovered video signal and *not* a DC level. The highest frequency results in an output of E_1 and the lowest frequency results in an output level of E_2.

It can be reasonably assumed that for a carrier which is frequency modulated the output level E will vary with respect to the carrier modulation. Sync tips which are the lower carrier frequency will be demodulated to the highest voltage level, the demodulated video signal will therefore be inverted (upside down). Further filtering removes the residual carrier leaving a clean inverted video signal to be corrected and buffered out.

Aperture correction

The term aperture correction may seem a little strange, but its history lies back in the earlier video camera equipment which suffered from defocusing

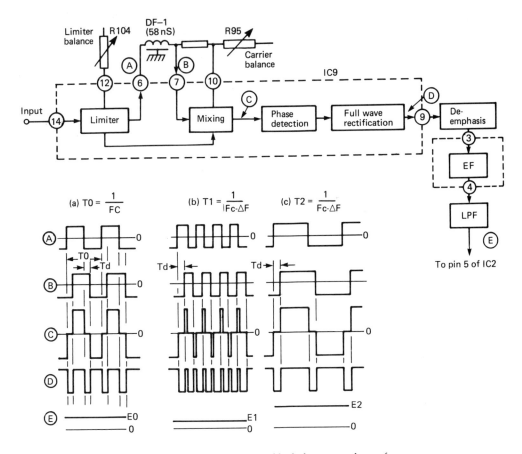

Figure 3.20. Demodulator block diagram and waveforms

of the picture when the iris or aperture was opened wide. The soft picture was corrected by crispening the edges to enhance it and this was called aperture correction. It is used in the replay path of video recorders to make a picture with a bandwidth of 2 MHz (−3 dB) look as though it had a bandwidth of 3 MHz :−10 dB) by enhancing the transient edges of the signals and effectively crispening the picture.

One aperture corrector is shown in Figure 3.21. The replayed video signal is split into two paths, one direct to a mixing amplifier and the other to a high pass filter. The input signal (A) is shown as a square wave; the transient edges have been softened by low bandwidth and they no longer have sharp rise times due to the high frequency losses. After passing through the high pass filter only the transient edges and noise are left as in waveform (B). These signals are then passed through a limiter to amplify and limit the transient peaks only. The limiter is designed to produce pulse type outputs from the duration of the input spikes. Following the limiter is a buffer driving a diode arrangement which removes the noise content as it is not of sufficient level to overcome the diode's forward voltage. A potentiometer taps off the required level of the pulses to the mixer input, and this is the aperture level. The transient pulses are then added back to the original waveform (A) to sharpen up the edges and to crispen up the signal as in (E). Note that (E) has very small overshoots and the original amount of noise. Adjustments of the aperture level control will increase, or decrease, the overshoots. This correction is sometimes referred to as edge enhancement and it will be found in some form or other in all domestic videorecorders; the Grundig 2 × 4 series has an external control for the user to adjust to taste.

37

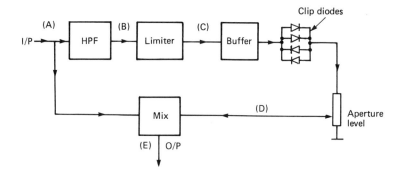

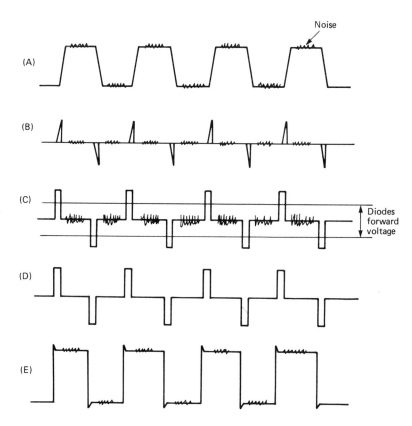

Figure 3.21. Aperture correction

A point worth mentioning here is that at times customers have stated that a replayed video picture is better than direct off air. This, of course, cannot be possible unless the customer's TV has a soft tube and the edge enhancement crispens up the picture. Apart from aperture correction most videorecorders have extra HF drive either in the UHF modulator or in the amplifier drive to the modulator, and a replay can in fact look better when the replay is via RF into the TV, as opposed to direct video replayed into a monitor. When a monitor is involved the author usually peaks up the aperture corrector, as the extra HF lift is not in circuit.

FM record and replay, Betamax variations

We have covered the basic VHS techniques for record and replay, and further advances will be covered later. Some of these advances in VHS technology are already available in some more popular Betamax videorecorders. The two main differences in Betamax are: white compression and a shift in the FM carrier for noise reduction. Figure 3.22 is a typical Betamax FM recording block diagram taken from a Toshiba V8600 but it is seen in Sony and Sanyo using the same IC but with some options not used in the more basic videorecorders.

The incoming signal is DC clamped and the sync tips are used for the clamp level as opposed to the back porch of the sync pulse, or black level. The reason for this is now more obvious, as the FM carrier is modulated from sync tip to peak white. The 'set carrier' potentiometer, R354, sets the DC level to which the sync tips are clamped, this is then the lower carrier deviation value of 3.8 MHz, and the signal is DC coupled through to the modulator. Some pre-emphasis is applied to the first stage at pin 13 and then further pre-emphasis is applied to pin 15 in the second stage. The pre-emphasis generates overshoots on the signal; black clip is a fixed level within the IC. White clip is adjustable and should be adjusted for about 55% to 60% overshoot; this can be slightly higher than in VHS due to the following white compression, but on the other hand the VHS double limiters can handle the overmodulations without black speckling after whites. Figure 3.23 illustrates the effect of the white compression network, with a linear staircase shown with the peak white overshoots caused by pre-emphasis. The top two steps are progressively shorter in height due to the compression action which also reduces the overshoots to a level which will not cause overmodulation.

A third stage, pin 15 of IC 401, has two functions: a non-linear pre-emphasis and an RF switching signal (the flip-flop). The non-linear pre-emphasis adds more pre-emphasis to the lower level high frequency components than it does to the higher level high frequency components, a sort of video 'Dolby'.

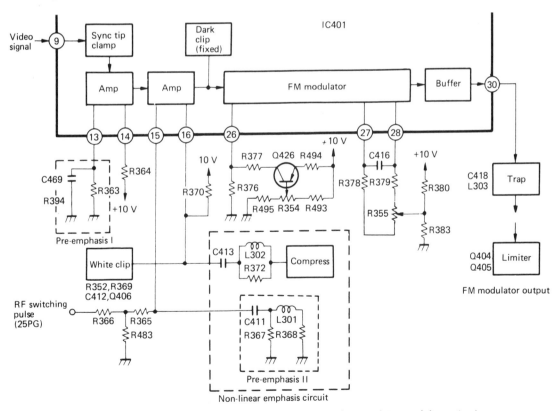

Figure 3.22. Betamax luminance non-linear pre-emphasis and FM modulator circuit

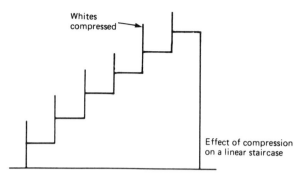

Figure 3.23. *Effect of compression on a linear staircase*

does. This switching signal has the effect of shifting the whole spectrum of the FM carrier by $\frac{1}{2}f_h$, which is 7.8 kHz. The FM carrier is shifted by $\frac{1}{2}f_h$ up and down on a head by head basis so each adjacent recorded magnetic track is displaced by $\frac{1}{2}f_h$ with respect to the one next to it.

R354 sets the free running or sync tip carrier frequency and R355 adjusts the FM modulator symmetry. A prior AGC stage, not shown, sets the video signal amplitude and, as such, the peak white frequency. An identical setting up procedure as described for the VHS section involving a frequency generator can be utilised to set the carrier deviation.

Figure 3.24 shows some of the replay correction, high frequency peaking and smear correction used to correct the signal after demodulation; the RF switching signal is used twice in replay. First it is used as a standard replay pre-amplifier head switching signal and secondly as an $\frac{1}{2}f_h$ restorer.

The $\frac{1}{2}f_h$ restorer is only to remove the up/down shift in black level caused by its application in the record mode; if it was left then there would be a change in brightness of the fields in replay and this would show up as a 'flashing' effect. The non-linear pre-emphasis is also corrected.

Although there are other Betamax recorders, earlier models, which do not have the non-linear de-emphasis, the difference is not usually obvious. In the most extreme cases black speckling after white captions will result if tapes are interchanged.

The RF switching signal has a more important function for noise reduction in the Beta system as we shall see later. It is important to understand at this point its effect upon the video signal and the FM modulator. The switching signal developed from the servos is high for one video head and low for the other and it is applied to pin 15 at a very low level, approximating to 5 mV p-p.

The switching signal modulates the DC level of the video signal up and down, the black level of the video signal is higher for Ch.1 video head than for Ch.2 video head. If the DC level of the video signal varies up and down for each video head then it follows that the FM carrier will also vary up and down for each video head and indeed so it

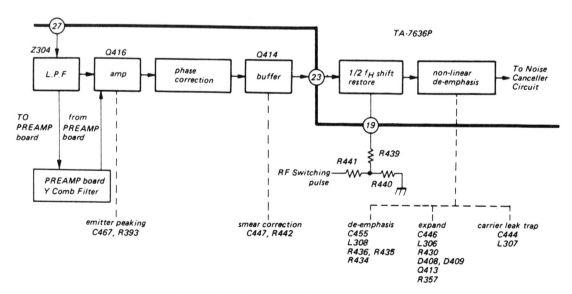

Figure 3.24. *Playback waveform shaper circuit*

Crosstalk cancellation

The crosstalk cancellation system devised for Betamax appeared in later machines in its useful format; the Sony C7 was first out with it and although the Rank Bush Murphy/Toshiba V5470 had the ICs for the system, it was not used until the following V8600 models. The clue is to look in the circuits of a given videorecorder for two demodulators.

In replay, as each video head scans its magnetic track it also picks up some FM carrier from an adjacent track. Now we all know about adjacent FM carrier rejection due to the azimuth differential in the video head gaps but the attenuation is not total and so some adjacent carrier is picked up. This carrier appears as crosstalk and is responsible for an effect that looks like symmetrically patterned noise, often referred to as an orange-peel effect, as it looks like the surface of an orange. This crosstalk of FM carrier will also contain other line frequency related noise components and all of this 'noise' can be reduced, if not eliminated, by a technique called 'Y comb filter'. The basis of this comb filter is to shift the recorded FM carrier by $\frac{1}{2}f_h$ (7.8 kHz), half line frequency, in record, so that each adjacent magnetic track has a half line frequency shift with respect to its neighbour (see Figure 3.25). At the top of the diagram is the replayed signal and noise spectrum; the black line

peaks are the replayed signal spectrum at line rate. The dotted line is the adjacent track crosstalk, because of the $\frac{1}{2}f_h$ shift in record the noise spectrum is interleaved between the wanted signal. If we then used a filter which had a spectral response like the teeth of a comb and the teeth were attenuators then it could be laid on top of the replay spectrum and eliminate the crosstalk peaks. Hence the name of comb filter, to interleave within the replay spectrum and attenuate the dotted line crosstalk. As you will also see in Chapter 5 the essential component of a line frequency comb filter is a one line delay line.

If we multiply out $\frac{1}{2}f_h$ we get

$$\text{crosstalk} = f_h \frac{(2N-1)}{2} \text{ where } N \text{ is an integer}$$

For any line N, where $N = 1$, crosstalk $= \frac{1}{2}$
where $N = 2$, crosstalk $= 1\frac{1}{2}$
where $N = 3$, crosstalk $= 2\frac{1}{2}$ etc

The meaning of these figures is that for any given line the crosstalk is shifted by a factor of $\frac{1}{2}$, which in terms of a carrier frequency is a shift of <u>180°</u>.

Figure 3.25(C) illustrates a carrier where $N = 3$, that is to say there are three whole cycles of the waveform, when it is compared to (D), which is a delayed waveform, delayed by one line. The two signals are in phase and this is the main wanted carrier.

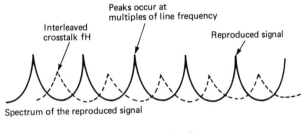

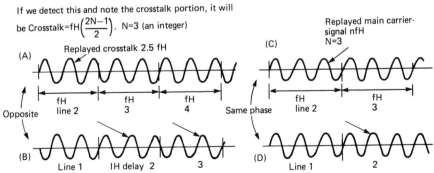

Figure 3.25. Principle of the Y comb filter

Figure 3.25(A) is the crosstalk, and for integer $N = 3$, the crosstalk = 2½. There are then 2½ cycles of crosstalk in lines 2, 3, 4 etc. This means that the odd half cycle in each line gives rise to 180° phase inversion between adjacent lines. If line 2 were then compared to delayed line 1, crosstalk in the two signals would be seen to be in antiphase.

The conclusion is that by shifting the carrier by $\frac{1}{2}f_h$ for each head in record then the crosstalk from an adjacent track in replay will be inverted in phase on a line by line basis; 180° phase shift.

Figure 3.26 illustrates the use of a delay line. In this example the wanted or main carrier components will add together and the crosstalk will cancel. Delayed line 1, added to direct line 2, will

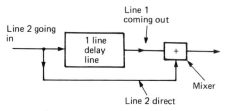

Figure 3.26.   Use of a delay line

contain 'in phase' main carrier and antiphase noise; note that the addition is carried out *after* demodulation so the main replayed carrier is also at a fairly low level, after passing through a low pass filter.

In the demodulator, the FM input carrier is squared in a limiter and it is used to injection lock an astable multivibrator. This runs as a delay of 90° from the input carrier. Figure 3.27 shows the demodulator and waveform diagram. The astable output and the limited input FM carrier are two inputs to a multiplier. The logic is that of an EXCLUSIVE-OR gate resulting in an output which is double in frequency to that of the input

carrier, similar to the demodulator described for VHS, again enabling higher carrier attenuation through low pass filters.

The constituents of the comb filter have been discussed along with the principles involved, it now remains to look at a practical working comb filter.

The block diagram of a working comb filter shown in Figure 3.28 is as used in a Toshiba V8600. In order to reduce manufacturing costs only a single 64 µs delay line is used, which doubles up as the delay line for the drop out compensator as well as the comb filter.

The drop out compensator is a standard technique. Replayed FM carrier is passed through a limiter and via a switch to the direct demodulator (1) and also to the delay line. A drop out detector, in the event of a drop out, switches the switch over and delayed carrier is fed to the direct demodulator.

The comb filter does not operate as such during a drop out, which is very short, so we will consider the comb filter as it is during a normal replay.

Demodulator (1) is in the direct replay path and demodulator (2) is in the delayed path out of the delay line; they are both identical and are followed by low pass filters. Each of the demodulator outputs contains demodulated video, residual carrier and crosstalk and are fed to a subtractive mixer (1). In mixer (1) the delayed signal is subtracted from the direct signal, video components and carrier components will be in phase and will cancel. Crosstalk is in antiphase on a line by line basis as previously shown and it will *add* inside a subtractive mixer. The output of mixer (1) will be crosstalk. Figure 3.29a shows the directly demodulated signal and 3.29b shows the delayed and demodulated signal. Note that the video signal is 'in phase' and that the crosstalk upon it is in antiphase. For clarity the residual carrier has been

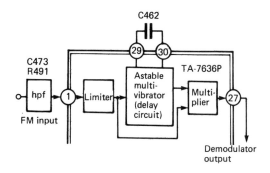

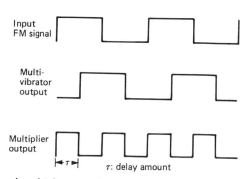

Figure 3.27.   FM demodulator circuit and multiplier waveforms

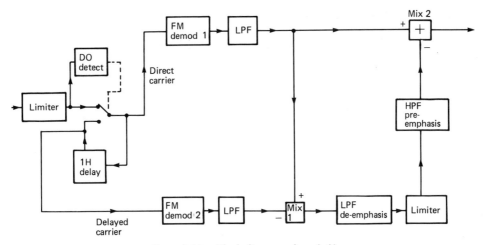

Figure 3.28. Block diagram of comb filter

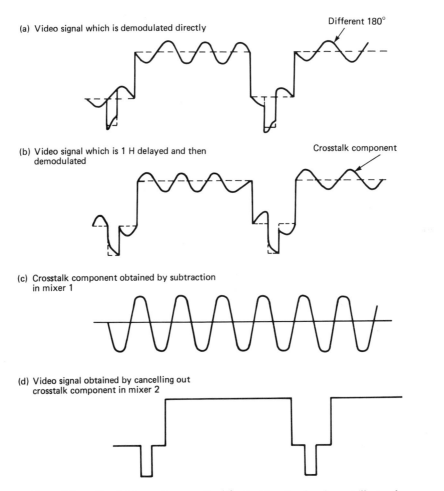

(a) Video signal which is demodulated directly

Different 180°

(b) Video signal which is 1 H delayed and then
demodulated

Crosstalk component

(c) Crosstalk component obtained by subtraction
in mixer 1

(d) Video signal obtained by cancelling out
crosstalk component in mixer 2

Figure 3.29. Comb filter noise canceller circuit video signal and crosstalk signal

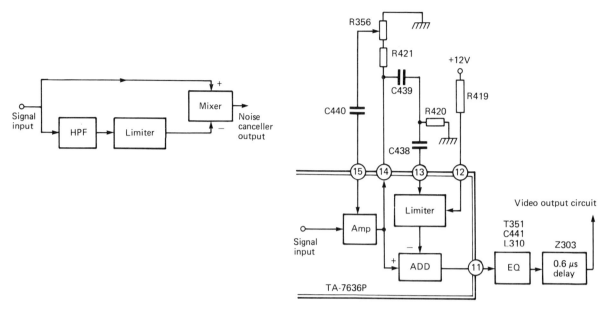

Figure 3.30. Noise canceller block diagram and circuit

omitted. When these two signals are subtracted the video cancels, leaving only crosstalk as shown in Figure 3.29c.

There is another problem, however. If any two consecutive TV lines are compared it will be seen that they do not contain the same signal content, indeed one could be a white signal and the next one, black. This would be the case in the event of a chess board replay. This would then be a difference component of the signal in the subtractive mixer and would appear at the output of mixer (1). If this video component in the crosstalk were allowed to be subtracted from the directly demodulated video in mixer (2), then a severe reduction in vertical resolution would result.

The output of the subtractive mixer (1) is then passed through a low pass filter, which is practical de-emphasis, to reduce transients, it then passes through a limiter, which clips the high level signal component to the same level as the crosstalk, and this will then reduce its adverse effect on vertical resolution.

The output from the limiter is then fed through a high pass filter to effectively pre-emphasise it again in order to subtract it from the directly demodulated signal which is still in a state of pre-emphasis.

Mixer (2) is then fed with two signals, the directly demodulated signal still containing crosstalk and the crosstalk signal with only a small amount of differential video content. Mixer (2) is also a subtractive mixer and the crosstalk input is subtracted from the directly demodulated input. In Figure 3.29 (c) is subtracted from (a) leaving (d) at the output, thus removing the crosstalk from the video signal.

There is still random noise left in the signal and this is removed in a further noise canceller as shown in Figure 3.30.

A subtractive mixer is used again and it is fed with direct video and the video signal fed through a high pass filter and limiter. The high pass filter is used to extract all of the high frequencies from the video signal, which will be both noise and HF transients, and the limiter clips the transients to the same level as the noise so that its output is basically noise. The noise is then subtracted from the video signal and cancels itself. R356 adjusts the amount of signal to the limiter and provides a range of adjustment from a soft picture to one that is very sharp and has overshoots and 'ghosting'. The delay line shown is just to equalise the luminance and chrominance delays in the signal processing.

FM recording and replay advances

In this chapter we have examined the principle of FM carrier interleaving. Its purpose is to minimise

44

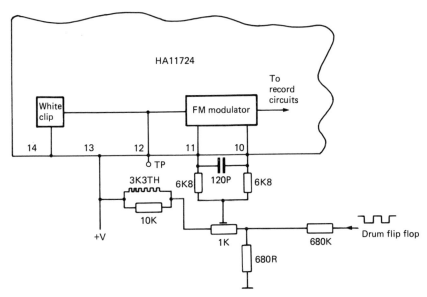

Figure 3.31. Carrier shift circuit

intertrack crosstalk at the relatively low carrier frequencies where the azimuth offset of the video head gaps is not fully effective. The idea is to shift the FM recording band by a factor of half line rate (0.5 fh, 7.8 KHz) on a field-by-field basis. The example given earlier was that of a Toshiba Beta machine; contemporary Sony models used similar arrangements.

During 1983 carrier interleaving techniques started to appear in VHS recorders, initially JVC model HR7600 and early-production HR7200; model HR7600 was soon superseded by the HR7650. Subsequent models in 1983/84 had the additional Long-Play facility, and the narrower tracks used here, with their lower signal-to-noise ratio made a more comprehensive noise reduction system even more necessary.

Inspection of the circuits of the *early* machines reveals that full carrier interleaving is not employed. Drum flip-flop signals are applied to the FM modulator to introduce a 7.8 KHz shift and the resulting interleaved carriers are recorded. During replay however no cancellation circuitry is employed, and it is left to the viewer's eye to interpolate the 'antiphase' crosstalk effects on successive lines. From May 1983 model HR7200 was fitted in production with an active cancelling circuit in the form of a sub-panel fitted beneath the tuning board; all versions of the HR7650 were so equipped. The only other VHS manufacturer employing the carrier interleaving system at that time was Panasonic.

Figure 3.31 illustrates the application of the 0.5 fh shift to the FM carriers. It is shown as part of a HA11724 FM modulator IC and can be identified in the circuitry of various makes of machines. The luminance signal, suitably clamped and clipped, is fed internally to the FM modulator – a test point is available on pin 12. Sync tip carrier frequency of 3.8 MHz is set by the 1 kΩ potentiometer and temperature-compensated against drift. The drum flip-flop squarewave is applied here via a potential divider of 1000:1 so its level at the FM modulator is very small: its effect is to increase FM oscillator frequency by 7.8 KHz for Ch.2 head while having no effect on Ch.1 head frequency. If a preset potentiometer is provided to adjust flip-flop squarewave amplitude do not adjust it without access to a spectrum analyser!

Replay correction techniques have varied somewhat as manufacturers have settled down to a standard IC or system. Figure 3.32 shows a typical VHS approach which utilises the existing drop out compensator delay line to additionally function in the noise cancellation circuit. In analysing this diagram it's important to remember that the 7.8 KHz shift introduced during record has the effect of arranging the crosstalk noise so that it is reversed in phase on each successive line during relay.

Referring to Figure 3.32, the replay FM signal level is stabilised by an AGC network before passing through the drop out switch to an equaliser. (Note that in later dual-speed recorders,

45

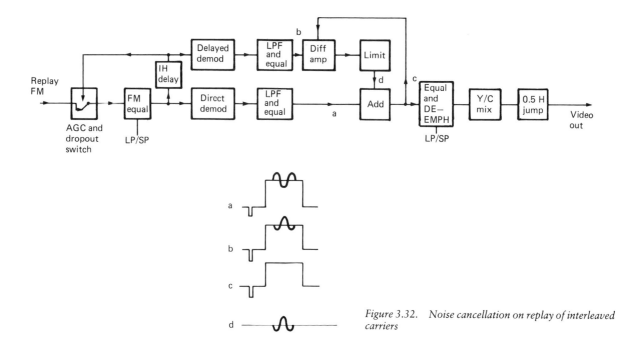

Figure 3.32. *Noise cancellation on replay of interleaved carriers*

Long Play/Standard Play compensation is provided at this point). After equalisation the signal takes two paths: to the demodulator and to the delay line. The delay line output feeds back to the drop out switch in the normal manner, and also feeds a second (*delayed*) demodulator in order to give two signal paths, direct and delayed. The main path is the direct one, in which the demodulated signal undergoes low-pass filtering and equalisation before passing (waveform a) into an adder. The one-line-old signal emerging from the 'delayed' demodulator is also filtered and equalised, then applied to a differential amplifier – waveform b. The differential amplifier's second input is waveform c, the 'cleaned up' output video signal which you will have to take on trust for the moment! Waveform b is subtracted from waveform c to leave just the spurious crosstalk component d which is amplified and amplitude-limited. The signal emerging from the differential amplifier contains not only the crosstalk-noise but also a large amount of unwanted video signal whose level must be reduced to that of the 'pure crosstalk' to prevent it from impairing the wanted direct signal excessively. Since waveform d consists of crosstalk *only* and is inverted with respect to the crosstalk riding on the 'direct' signal, the two waveforms a and d cancel in the adder, whose output now consists of pure video signal. This technique is only effective for low frequency crosstalk; HF components are dealt with in the next block, equalisation and de-emphasis, as shown and described in Figure 3.30 and p. 39. Where the circuit of Figure 3.32 is used in a dual speed machine a 0.5 fh jump circuit is incorporated in the signal output path.

A later development on this theme (Autumn 1984) is given in Figure 3.33, Here the basic system is deceptively simple – the circuit is in fact very advanced. After AGC and equalisation the FM carrier passes through a sophisticated double limiter before demodulation to baseband composite video.

Let us first examine the drop out compensator. This operates at *baseband*, taking a signal from the demodulator via the NC (Normally-Closed) drop out switch to the adder and to the delay system. The DOC changeover switch is driven by pulses from the drop out detector in the FM AGC section, so that whenever a break in carrier occurs the switch will toggle to 'patch' the hole with a one-line-old delayed signal. You may have noticed that the delay line operates on baseband video rather than at FM carrier frequency. This cannot be done in a conventional glass delay line, and the device used here is a CCD (Charge Coupled Device) 'bucket brigade' IC, to be described in a moment.

Noise cancellation for the frequency-interleaved FM carrier also takes place via the delay IC. The

46

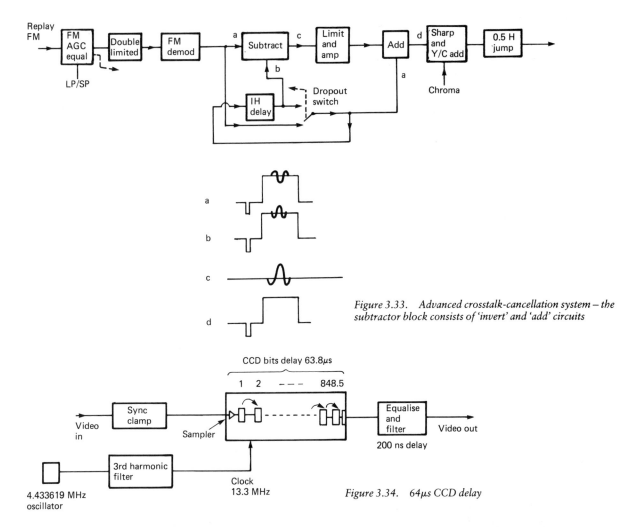

Figure 3.33. *Advanced crosstalk-cancellation system – the subtractor block consists of 'invert' and 'add' circuits*

Figure 3.34. *64µs CCD delay*

'direct' baseband signal (a) and the delayed baseband signal (b) are applied to a subtractor whose output then consists of pure crosstalk, c. This is amplified and limited for application to an adder, wherein the crosstalk component in the main signal path is cancelled to render the clean composite video signal d. The circuit incorporated in the replay path also reduces tape noise, rendering further noise reduction artifices unnecessary. After addition of chroma in the Y–C add matrix and edge-sharpening the CVBS signal is output via the 0.5 fh jump circuit, used in LP video search, see Chapter 8.

CCD delay line

In this context a Charge Coupled Device is an IC with a certain number of CCD packages, or 'buckets' within it, see Figure 3.34. The analogue input signal is sampled by a clock-controlled gate and the sample level at each clock pulse is passed to the first 'bucket'. On each subsequent clock pulse the analogue sample is passed on from each bucket to the next on the right until it arrives at the output terminal. In this way the signal is broken down into quantised units, passed along the bucket brigade within the chip at a rate dependent on clock frequency, then reassembled at the output end.

In the 64 µs delay CCD there are 848.5 buckets within the IC, and it is clocked at 13.3 MHz; clock frequency is derived as the 3rd harmonic of the 4.433619 MHz colour crystal oscillator. The clock period is thus 75 ns. 0.075 us × 848.5 = 63.8 µs. A further delay of 200 ns takes place in the filter/equaliser at the CCD output to make up 64 µs, exactly the period of one line.

47

VHS HQ

The addition of new technology to VHS has allowed the system to develop further to produce high resolution pictures. This is achieved in three main ways and any manufacturer adopting at least two of them may describe their product as HQ.

The first method to improve resolution is by increasing the white clip level. In videorecorders described earlier in this chapter the white clip level as given in the service manuals was set to 60% above white level, given that sync tip to white level was 100%. In VHS HQ machines this is increased to 80% so the overshoot is 80% of the overall video signal level (i.e. set to a limit of 180%). You will know from this chapter that the white clip is set to prevent overshoots created by pre-emphasis from driving the FM modulator above 4.8 MHz and causing replay inversion (black streaks on white lettering). In order to allow the new peak white level of 80% without overmodulation a new compression circuit is incorporated. It is included after the pre-emphasis circuits, or as a nonlinear control signal to the main pre-emphasis onto which the peak white clip is connected. The compressed overshoots are then expanded in replay so that the effect is similar to recording and replaying the video signal with peak white levels in excess of a maximum modulation frequency of 4.8 MHz without actually exceeding the maximum frequency.

The second method is to improve the detail of the picture by enhancing low level high frequency components and transitions from black to white and vice versa before the signal is recorded. In some recorders this is referred to as Dynamic Aperture Correction (DYAC).

The third method is to use the Y comb filter noise reduction system described earlier more effectively in a process called luminance line interpolation.

Dynamic aperture correction

Dynamic aperture correction is applied to the Y recording signal between the AGC circuits and the FM modulator circuits. Tied in with it is the non-linear emphasis and white clip circuits.

Figure 3.35a shows the block diagram of the system, formed by a single IC with the exception of equaliser EQ2. Figure 3.35b shows waveforms, exaggerated for clarity.

The aim is to emphasise the transient edges of low level high frequency components before recording, such that on replay a signal more closely approximating to the original is obtained. In previous models, edge recovery in replay has increased picture detail but at the expense of added edge noise: with DYAC pre-emphasis the same replay correction can be achieved — without edge noise. Note: edge noise shows up as busy little sparkles on all detail edges.

Waveform (a) is a low level square wave applied to the DYAC circuit and is split three ways: into a mixer, into equaliser EQ2 and into a low pass filter. The filter is not terminated at its output, and this will cause input signals to reflect back, effectively creating a 'ringing' delay line of period 't' where t=100 ns. Signals at its output (d) are delayed by 't' whereas reflected signals back to the input (b) are delayed by 2t. The input to the first mixer (c) is, therefore, the sum of input (a) and reflected signal (b).

Output of the low pass filter (d) is inverted and added to (c) by the mixer, so we have (c) - (d) = (e).

Mixer output (e) is processed by a noise canceller to remove noise present during high input levels to prevent over-compensation, as the DYAC operates on low signal levels and the circuits have to ensure that high noise levels are not considered to be low level video.

After peak white and dark overshoot clip levels the amount of dynamic aperture correction is controlled by adjusting the gain of the inverted phase of signal (e) before it is added to (f).

In practice the signal (e) undergoes a delay of $\frac{1}{2}$ in the signal processing, making a total delay of $1\frac{1}{2}t$ and, in order to compensate for this delay, signal (f) is passed through an equalising delay of $1\frac{1}{2}t$ (150 ns) in EQ2 thus ensuring (e) and (f) are coincident. The strange looking result (g) is a signal which has pre-compensation. When de-emphasis is applied in replay the picture will look sharp and will not have the 'busy' looking edge noise that was so obvious in earlier models.

Luminance line interpolation

We have discussed the effect of introducing a $\frac{1}{2}f_h$ shift into the FM carrier to reduce crosstalk from the lower frequency levels of adjacent track FM carriers by the use of a Y comb filter and this is now part of the VHS HQ specification. Further reduction of high frequency tape noise can be achieved by the extended use of the Y comb filter.

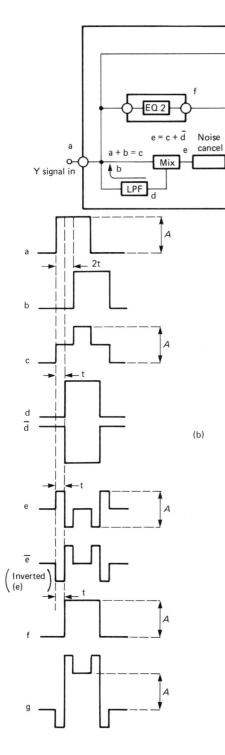

Figure 3.35. a, dynamic aperture CTL block diagram. b, DYAC timing chart

By the addition of two consecutive TV lines, increasing the signal by a factor of 2 while decreasing noise by a factor of $\sqrt{2}$ as shown in Figure 3.36.

As noise correlation exists between consecutive lines further reduction can be achieved by extending the principle over a number of lines, but adding previous lines on a reducing level basis. The reduction in noise, then, is:

$$20 \log \sqrt{\frac{1-k}{2}}$$

where k is the reduction in level of previous lines, and is 0.5 in VHS HQ.

In Figure 3.37 the first signal is n at full level, n-1 is added to it but at a level reduction of k = 0.5, n-2 is then added but reduced even further by another multiple of 0.5, hence $0.5 \times 0.5 = 0.25$ of n. Therefore when k=0.5 each delay line reduces the signal by half, so over a number of lines 'n' the reduction of the added signal is k^n.

The practical approach is to use a circulatory delay line system shown in Figure 3.38 where the output of the delay line is reduced by ½ each pass: equivalent to k=0.5. Each line is added into the circulatory system and each time it goes around it is reduced by k, eventually to an insignificant level, so a number of previous lines are being added to the input signal each of a reducing level. The overall effect is that the video components are

49

additive and the random tape noise is subtractive, effectively forming a circulatory filter. The output of the circulatory filter is a video signal with noise and high frequency components removed, it is also increased in level and so it is reduced in level by the (1-k) network to the same as the input video.

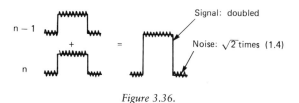

Figure 3.36.

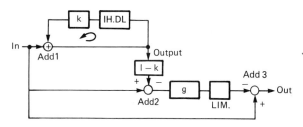

Figure 3.37.

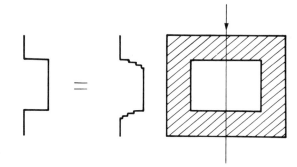

Figure 3.38. Playback

In Add2 the 'clean' output of the circulatory filter is subtracted from the input video leaving a resultant signal which is mainly noise and transient video components. As the video transients will have adverse effects they are removed by amplification and limiting leaving a signal which is a sum of random noise. Lastly the noise is subtracted from the main video signal within Add3 thereby cleaning it.

There is a problem in that this circulatory system would reduce vertical resolution as shown in Figure 3.39 due to the addition of adjacent lines, especially when one line is black and the next contains a full white. In order to minimize loss of vertical resolution in playback the circulatory delay line is brought into use during recording as a vertical pre-emphasis circuit. The pre-emphasis system is shown in Figure 3.40, where the input signal is shown as a TV field signal with line sync pulses, indicating a white square on a black background. The circulatory system along

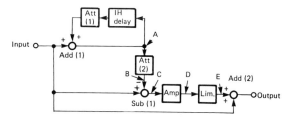

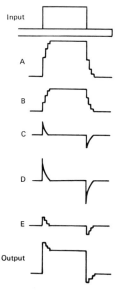

Figure 3.40. Record

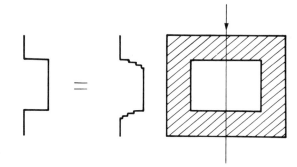

Figure 3.39.

with ADD1 produces the signal (A) with poor vertical resolution and increased level. This is reduced by attenuator ATT2 to the same level (B) as the input signal from which it is subtracted to result in waveform (C). (C) is amplified to (D) and limited to give (E) which is then added to the input wave to produce the pre-emphasised output.

In playback (Figure 3.41), the replay signal (note that its overall shape is not affected by the record/replay process) contains noise. Again the circulatory system produces an increased signal level with the noise removed (A) which is attenuated to (B) and subtracted from the input signal resulting in (C). (C) is then attenuated slightly (D) and limited to give (E) which is then subtracted from the input signal resulting in an output signal closely resembling that of the original input signal.

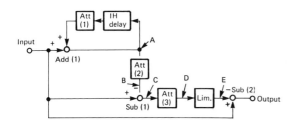

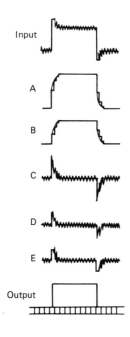

Figure 3.41.

Super VHS

The specification for Super VHS (S-VHS) pushes the luminance band even further up the scale to a carrier frequency between 5.4 MHz to 7 MHz giving a massive increase of resolution from 250 to 400 TV lines. The deviation is 1.6 MHz also improving the video signal to noise ratio. The spectrum is shown in Figure 3.11g.

The bandwidth of the colour signal is enhanced slightly by increasing the bandwidth of the downconverted recording signal: it is centered on 627 kHz with a bandwidth in excess of 500 kHz. The luminance white and dark clip levels have also been increased along with the system's capability to handle them: white clip is 250% and dark clip 70%.

The new system specification also reduces cross colour effects and this is illustrated in Figure 3.11g, by tailoring the bottom end of the luminance carrier and the top end of the colour carrier so that the two do not overlap, leaving a gap of 200 kHz–300 kHz. There is a more sinister effect of this though which will no doubt cause upset in the Scart connector camp: JVC is constructing new S-VHS machines with separate luminance and chrominance inputs and outputs, in order to fully utilise the cross colour and bandwidth improvements.

New tape

A very high grade tape which is intended for S-VHS has been developed as an exclusive video tape, employing a new highly efficient cobalt iron oxide material. An identification hole is provided in the cassette for the video recorder to identify the type of tape.

Compared with existing high grade tapes and super high grade tapes which have high coercivities of 720–750 oersted, the S-VHS tape has a coercivity of around 800–900 oersted.

Compatibility

The S-VHS recorders are dual VHS standard and will record and playback to both system specifications, however previous models will not playback the S-VHS specification tapes.

LSI video processor for luminance record and playback

This is not in fact an IC but a small PCB which contains the processing LSI integrated circuit. A feature of this kind of circuit is that it embodies a number of processing functions for the top of the range videorecorders which are left dormant or wired out for lower cost versions or not used to full capacity. For fault finding it can get confusing if the repairer is not careful as quite a number of the signal paths are bi-directional and used for both recording and playback, switching is carried out within the IC.

Input video is applied to pin 3 to the keyed internal AGC circuit, provision is left on pin 1 for E/E level control but not normally used. With the switch in the E/E or record mode the video signal

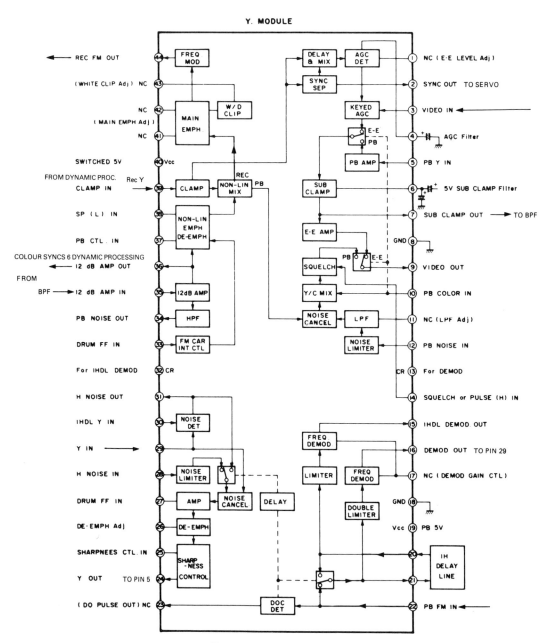

Figure 3.42. Y module block

is clamped and amplified and then appears on pin 9 as the monitor E/E signal. Meanwhile the recording video signal comes out of pin 7 and goes to a band pass filter to remove colour and then back into pin 35 for amplification between pins 35 and 36 to compensate for the loss in the filter. From pin 36 the signal is split, one path will take the filtered Y signal to the colour processing for sync separation to derive the H syncs for colour AFC loops.

A second path is Y processing, that is 'dynamic aperture' and 'record Y circulatory pre-emphasis' prerecord shaping of the signal, these are optional extra circuits which may not always be present. A third path uses the Y signal for non-linear pre-emphasis attached to pins 37/38 to add back to the main Y signal when it returns via pin 39. Pin 39 is where the Y signal returns to the IC and it is again clamped, a reference for the AGC detector is used here, this means that all of the Y processing so far is also within the AGC loop and compensates for tolerance variations within the whole path up to this point. As the level of the Y signal is stabilised a sync separator can be used for timing the servo circuits making the servo syncs independent of video level changes.

Non-linear pre-emphasis is added to the Y signal along with DC level changes from the drum flip flop signal (pin 33), on a dual speed version SP/LP control of the amount of pre-emphasis is on pin 38. Pin 37 has playback control for reversing the roles of circuits for de-emphasis and DC level flip flop cleaning (to prevent replay flicker) in replay. If you are not sure of the role of the flip flop signal at this point refer to the sections dealing with Y carrier $\frac{1}{2}f_h$ frequency shifting.

Further main emphasis is added between pins 41 and 42 with external control, but this control is not normally used. White clip is applied to pin 43 prior to frequency modulation and dark clip is fixed within the IC. This is due to the stable DC or black level of the signal within the confines of the IC. White clip level is also semi fixed but adjustment is catered for on pin 43. Record FM carrier is out of pin 44 to the recording drive circuits.

Replay

Replayed FM carrier enters at pin 22, and is processed in the cyclic drop out compensator, two outputs are derived, direct (pin 16) and delayed (pin 15). The main output is on pin 16 which is double limited and demodulated ready for noise reduction processing. The delayed and demodulated output on pin 15 is for the Y signal carrier interleave noise reduction circuits, this may not be used on lower cost versions. Demodulated Y signal from pin 16 is routed back to pin 29 to the noise detector, limiter and canceller. H noise is on pin 28 and noise detector sections on pins 30 and 31 are for filtered HF noise for standard noise reduction in lower cost machines and is an optional section of the IC circuits.

Replayed Y signal is amplified (27) and de-emphasised (26) before passing to the sharpness circuits with control on pin 25, de-emphasis adjustment on pin 26 is not normally used. The drop out pulse on pin 23 is used to switch off the noise cancelling circuits during the insertion of signal in the event of a 'drop out' when their operation is counter productive.

From pin 24 out of the sharpness circuit of the IC the Y signal re-enters at pin 5, play back Y in, with the switching now in the PB position. After clamping the signal is passed out of pin 7 to the circulatory noise reduction system, now also set to playback, it re-enters the IC at pin 39. Non-linear de-emphasis is applied and the signal routed internally to HF noise cancellation circuits before replayed colour is added. The output composite video signal is on pin 9 after passing through the squelch circuits. Squelch 'high' is present during muting and synthesised vertical sync pulse insertion during cue, review and still picture modes of operation.

This is a typical example of multi-role signal processing within a single unit capable of working a wide range of recorders from basic to high quality. The fidelity of video replay pictures is dependent on the type and number of noise reduction systems that are used within the confines of head/tape bandwidth.

4

SERVOMECHANISMS

Servomechanisms are appearing as standard features in domestic equipment more frequently. This is due to the cost effectiveness of large scale integrated circuits to provide close tolerance speed control in such applications as record player turntables, audio cassette recorders, disc players and of course video recorders.

Commonly, video recorders have two servos, one to control the video head drum and the other to control the tape speed, the latter being the capstan servo. The dynamic track finding principle used by Grundig/Philips in the 2000 series recorders may be considered to be a third servo mechanism, as it acts upon the video heads.

Any servo is a three terminal device, with two inputs and an output. In video recorders there is a steady reference input and a positional feedback input, which are acted upon by the servo electronics to provide an error correcting output, which may be a drive voltage with the error signal carried upon it, or it may take the form of a control signal.

The servomechanisms shown in Figure 4.1 are the very basic sort. The left-hand one is a servo with direct drive to a DC motor, PG (pulse generator) pulses are developed from small magnets attached to the circumference of a rotating flywheel. These PG pulses are compared with the incoming reference signals, usually incoming video. An error voltage is derived and added to, or subtracted from the motor drive voltage to speed it

up or slow it down. This keeps the motor going at a speed dependent on the frequency of the incoming reference.

The right-hand diagram also shows a servo in which the motor is still held to a speed dependent on the reference, however this time the motor is an AC synchronous motor. The speed of rotation of the flywheel is chosen to run at a speed just a few cycles per second higher than the servo speed. The output from the servo is a control voltage to a magnetic brake, an eddy current brake, which slows down the motor by loading it with a magnetic field. Once the correct or 'locked-in' speed is reached the motor is controlled by varying the current through the eddy current brake coil, thus increasing or decreasing the loading effect. The eddy current is a result of the comparison between incoming reference and feedback PG pulses.

Note that there is a 'phase' inversion between the two methods. An increase in motor supply voltage will speed it up, whereas an increase in eddy current will slow the motor down.

The head drum servo

The objective of the head drum servo is to ensure that during recording the recorded video is put onto the tape in a track pattern that conforms to

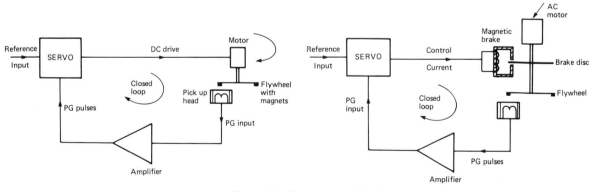

Figure 4.1. Basic servomechanisms

the standard laid down for that particular recorder. It uses the incoming video as a reference and a PG signal from the head drum as motional feedback. This is in order to run the drum at the required speed *and* line the video heads up to such a position that they can record each slant track as one field of video. There are two heads on the drum, spaced apart by 180°, and each records one field of video information. The tape is wrapped more than 180° around the drum so that each head records slightly more information than is required for replay. This overlap of information in record is necessary to provide for continuous FM signal during replay.

Incoming field rate is 50 Hz, there are two heads on the drum cylinder and each records one field. It therefore follows that two fields, that is one 'frame', are recorded each revolution. The drum speed is half of the field rate, that is 25 Hz, equivalent to 1500 rev/minute or a time period of 40 ms (milliseconds) per revolution. The frequency of operation of the servo is 25 Hz or a period of 40 ms. Servo timing signals are measured in milliseconds and are normally quoted this way as you will see.

The video tracks have to be recorded accurately, so the position of the video heads around the circumference of the drum, when it is running at the correct speed, is controlled by the servo with

reference to the incoming video. This is called 'phase control'.

Figure 4.2 shows a typical track pattern. Two video tracks for head Ch.1 and head Ch.2 are shown as recorded fields. A and B are head switching points where the heads change over during replay. You will note that there is information recorded either side of the crossover switching points to allow for overlap, due to tolerances. The switching points are normally about 6½ lines before field syncs. The field syncs must be recorded at the same distance up each track, otherwise picture jitter would occur. Having now determined that each head has to be at a particular position when travelling up the video track at the time field syncs appear, it follows that the position of the video heads has to be related to the video field sync. A steady relationship is developed between the incoming video and the position of the video heads around the drum at a speed of 1500 rev/min. This is done by comparing field syncs with the PG pulses from the rotating drum, or its flywheel. The PG pulses also provide the head switching signal, which on replay will select each head in turn, such that when the outputs are mixed a continuous signal is recovered.

If we look at Ch.1 head, in Figure 4.2, scanning up the tape from crossover point A1 towards B1, having already recorded field syncs. When it reaches point B1 and is about to leave the tape, Ch.2 head has already contacted the lower edge of the tape at point A2. On Ch.1 track after point B1 Ch.1 head records a small amount of overlap, similarly on Ch.2 track, Ch.2 head records a small amount of excess signal prior to point A2. These two sections of extra signal at the end of one track and the beginning of the next are duplicate recordings and carry the same signal. In replay, switching Ch.1 head off and Ch.2 head on during these duplicated signal periods will result in the replay signal being continuous, without breaks. The time period between the last line sync pulse replayed by Ch.1 head and the first one replayed by Ch.2 head will be 64 us. This is called horizontal correlation; maintaining the correct timing of line sync pulses over head crossover points.

Figure 4.3 shows the replay from each head, the switching signal developed from the PG pulses and the resultant output of FM carrier from the pre-amplifier circuits. It can be seen that each head replays more signal than is required, so the switching signal from the PG pulses is delayed as necessary to arrive at the time when one head is about

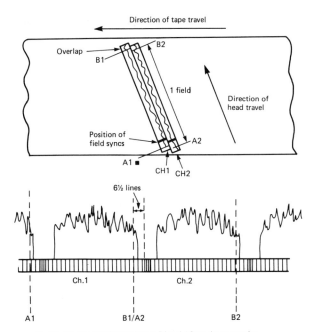

A1, B1, A2, B2 are switching points, B1 and A2 are the same point

Figure 4.2. Typical track patterns

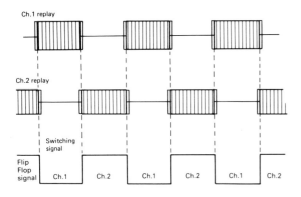

Figure 4.3.

to leave the tape and the other has just started scanning. During the time that the heads are overlapped they are duplicating the recorded signal so that switching from one head to the other does not cause any discontinuity. A small disturbance of about two lines is visible, but as this is 6½ lines before field syncs it is placed at the bottom of the picture out of sight.

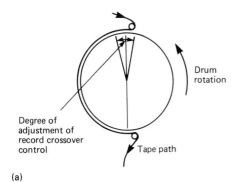

(a)

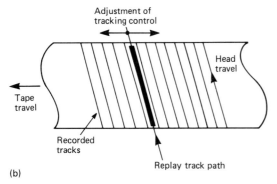

(b)

Figure 4.4.

The drum revolves at 1500 rev/min or 25 rev/s during record or playback. In order to record the video tracks accurately the video heads have to be positioned at a certain point on the circumference of the drum. Phase control is used to hold the video heads at this position so that the heads start to contact the tape some 12–15 lines before field syncs occur, that is the overlap period before head switching. The heads switch only on replay but that does not alter the fact that the changeover on record has to be defined. There are tolerances in any system so some adjustment is required during the recording to align the heads to the correct point. This is done by comparing the incoming video against the head switching signal, referred to the drum Flip Flop or RF switching signal and adjusting the 'record switching point' control so that one of the edges coincides with a point on the video signal which is 6½ lines before field syncs. The adjustment is usually called 'record crossover position'. Figure 4.4a shows the approximate range of control. In replay the 'tracking control' operates in much the same way but the degree of control is greater.

When the tape is replayed each head will replay its own track but it may not be centralised on the middle of that track; it may be one side or the other, so it is necessary to make some provision for adjustment. The adjustment is the 'tracking control'.

The tracking control can operate on either the drum servo or the capstan servo. During record on the drum servo a small amount of adjustment is provided to alter the position of the video heads, about 15° around the circumference of the rotating drum. In replay the tracking control is similar but with a greater range, enabling the heads to be shifted by a whole replayed track, that is equivalent to 180° of phase adjustment.

A typical drum servo

A block diagram (Figure 4.5) and a timing diagram (Figure 4.6) will illustrate how a servo can work, using phase control to achieve the objective of video head positioning. Magnets M1 and M2 are placed on the drum, or its flywheel, in such a position that M2 produces an output just after Ch.1 head has recorded field syncs. This relationship is deliberate, as a variable delay can then be put in the path of field syncs to derive a sample pulse, the position of which can be altered. There is also a divide by two circuit. The servo will be

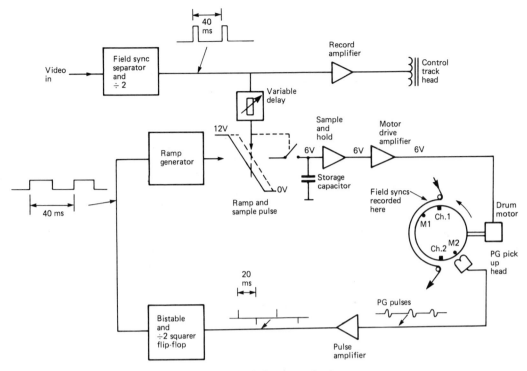

Figure 4.5. Block diagrams of a drum servo

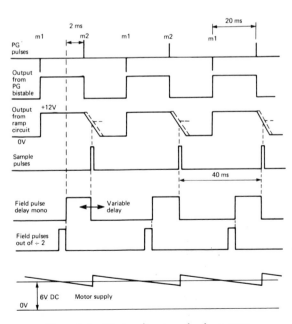

Figure 4.6. Timing diagram of a drum servo

running at a period of 40 ms, so the 20 ms field pulses are counted by two, or alternatively, a monostable with a delay of about 30 ms will eliminate every other field pulse. Control track pulses are also recorded at a period of 40 ms for use during replay.

Picking up the loop operation at an output from magnet M2 giving a PG pulse, this is amplified to a suitable level in a pulse amplifier and then passed on to a bistable, which acts as a divide by two and squaring circuit. A negative transition out of the bistable is slowed up in a ramp generator circuit. M1 causes positive edges on the bistable and M2 negative edges, indicating which head is where. The bistable output, the flip-flop signal, is low for Ch. 1 head and high for Ch. 2 head. A sample pulse from delayed field syncs arrives at a point which coincides with halfway down the ramp. If the ramp has a voltage level of 12 V then halfway down will be 6 V. The following circuit is a 'sample and hold' system and acts like a switch, which for a short period of time is closed by the sample pulse. The 'switch' is closed for only the short duration of the sample pulse, but it enables the instantaneous ramp voltage to be transferred to the storage capacitor.

When the switch is opened again the capacitor cannot discharge except through a high impedance amplifier. As it is topped up every 40 ms a 6 V DC voltage is formed across it. We are then sampling instantaneous ramp voltages, provided the sample pulse arrives at the same time with respect to the ramp then the voltage will remain at 6 V. Voltages stored across the capacitor are buffered and used to drive the DC motor, and the motor has been designed to run at the chosen speed at 6 V. The head drum will therefore run at the correct speed as long as the sample and hold voltage remains at 6 V.

The sample switching pulse is derived from the incoming video via the short delay which is variable. This delay is the time difference between the pulse from M2 and the position of the video head when recording the field sync part of the incoming video waveform. By varying the monostable delay time of the sample pulse we can shift the point on the recorded video track where field syncs are recorded and hence the record crossover point.

If during an operation the head drum was slowed down very slightly, due say to an increase in tape tension causing friction, then the M2 pulse may arrive late, shown as a dotted line to the right on the timing diagram. This would make the ramp late with respect to the sample pulse, which would then sample higher up the ramp, and the sample and hold DC voltage would rise.

If then the sample and hold voltage increases, the drive voltage to the drum motor will also increase. The motor would then have extra power to overcome the friction which tried to slow it down. The correction control of the drum servo is continuous control so in real life the motor does not actually slow down, a slight tendency to do so is immediately compensated for. The servo corrects for any undesirable effects without altering the relative position of the video heads with respect to incoming video.

The relationship between M2 and the position of the video head is determined by the position of M2 on the drum (or flywheel) and the short delay on incoming field pulses. That is to say that when M2 passes the pick up head and generates a PG pulse, the video head is on its way up the magnetic track having just recorded field syncs. This is supported by the timing diagram which illustrates that field pulses, 40 ms apart, occur about 2 ms before the M2 PG pulse.

The drum servo is designed to maintain this timed relationship, therefore it ensures that the field syncs are recorded at their correct position on the magnetic track every time. The manufacturer's track standard is maintained and interchangeability is possible within the given standard.

Phase locked loop operation

It is much easier for TV engineers to consider the operation of a drum servo if it is compared to that of a line timebase. Within a line timebase you have an incoming reference signal, line sync pulses. These pulses are compared to feedback from the line oscillator via the line output transformer. In Figure 4.5 let the line oscillator and LOPT be analogous to the motor and the sample and hold circuit, to the line phase discriminator. Incoming line syncs are compared to feedback from the output stages in a phase discriminator, which is fed with a sawtooth signal (ramp) from the LOPT. If the line oscillator is running at the correct frequency (speed) then the time base will lock, as does a servo.

Once the line timebase is locked you can shift the picture sideways on the screen by slightly adjusting the line hold or the line oscillator coil; too much adjustment and you will lose lock. The sideways shift is due to the introduction of a phase shift between the incoming line syncs and the timebase feedback loop. Phase advance will shift the picture left and phase delay will shift it to the right.

This is exactly the same as the drum servo, when it is locked. If you strobed the drum you would see the video heads in a fixed position on the circumference of the cylinder. It can be done under the influence of a fluorescent light, but a slight drift will be evident as the fluo and the servo will not be synchronous! Adjustment of the 'record crossover position' control will shift the position of the strobed heads clockwise or anticlockwise, within the limits shown in Figure 4.4a. Too much adjustment would cause the servo to lose lock, however the range of the control in practice is limited. Clockwise shift is phase advancement of the motor feedback loop to incoming field syncs. It is achieved by reducing the timing period of the variable delay, monostable. Obviously, increasing the mono time period will create a phase delay and move the video heads anticlockwise.

You should now be able to understand the phase relationship between incoming field syncs and the rotating video heads, and how a shift in that phase will determine the position of the field syncs within the recorded tracks. In playback a shift of 180 degrees will 'change the heads over'.

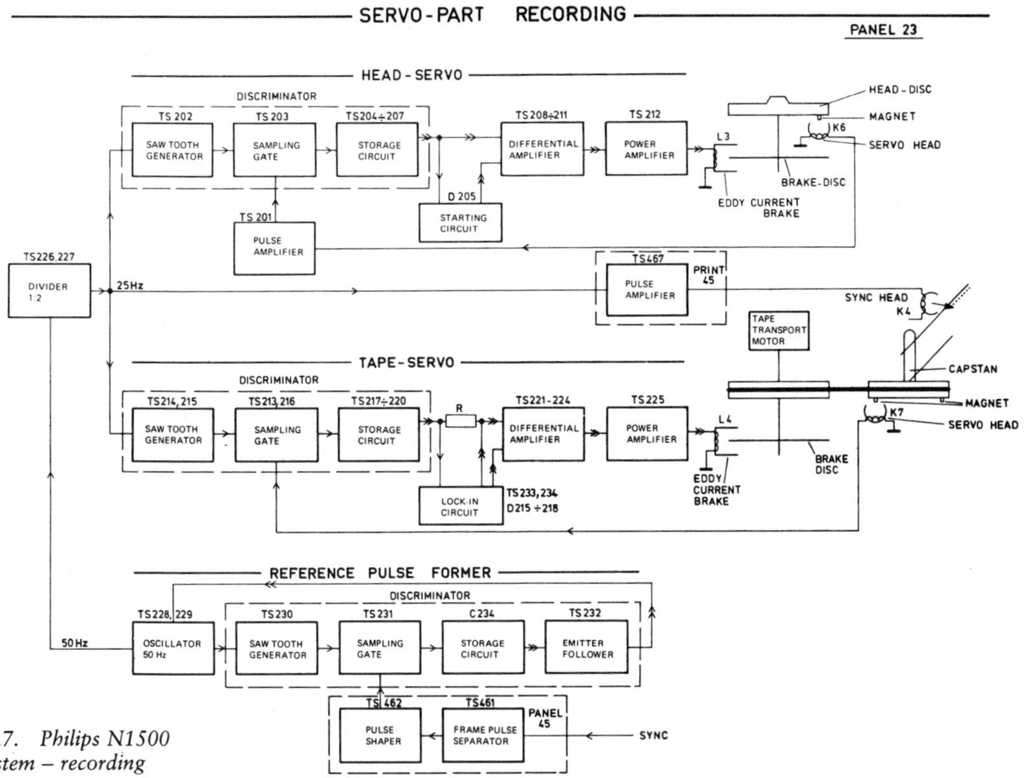

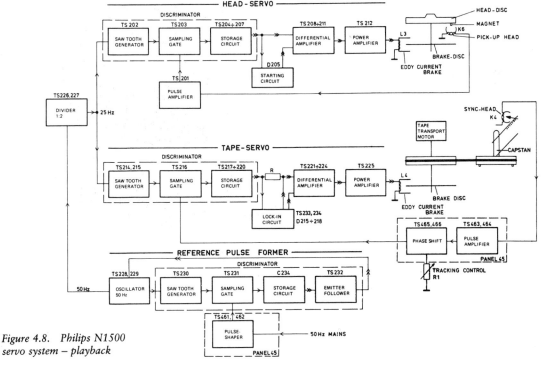

Figure 4.7. Philips N1500
servo system – recording

Figure 4.8. Philips N1500
servo system – playback

59

MISC	K7		D214	TS213	TS227	D212	D211		TS226	D210		TS214		TS215			D207	TS216	TS218	TS217	D208
	K6	TS228		TS201		TS229	D213		TS230		TS231	D201		TS202	D202		D203	TS203	TS232	TS205	TS204	D204
C		212					224÷227					214		216				203				
		228	201	202	230	229			231		232		233		204	205	236 240					206
R			241	288	307	283	285÷287		284		278÷282	238	242		243÷246			247÷249			250 251	
		201÷204	289÷295							296÷298		299÷301	205÷207 302				208÷211	303÷306		212 213	
MP	201−209 219								225		224	226	218				221 220			227		229

CIRCUIT DIAGRAM D

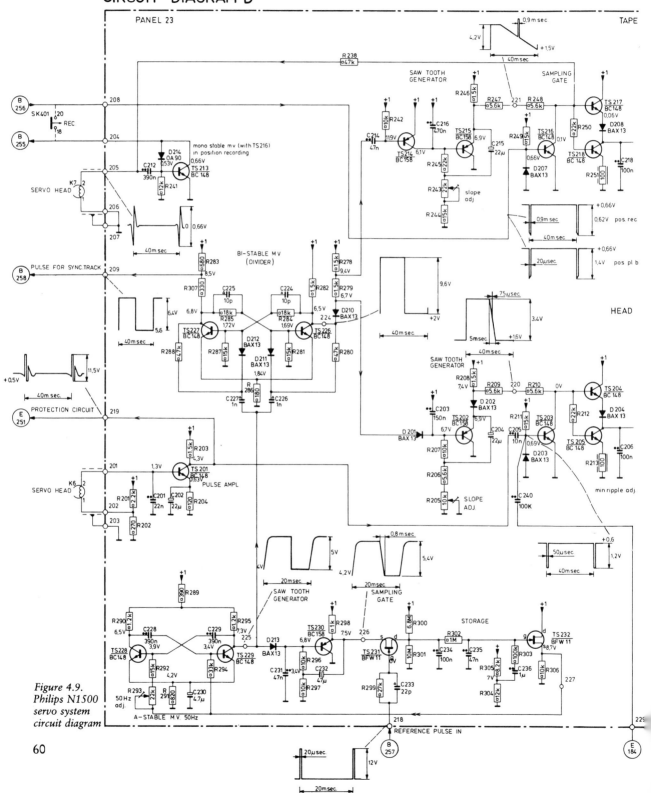

Figure 4.9.
Philips N1500
servo system
circuit diagram

60

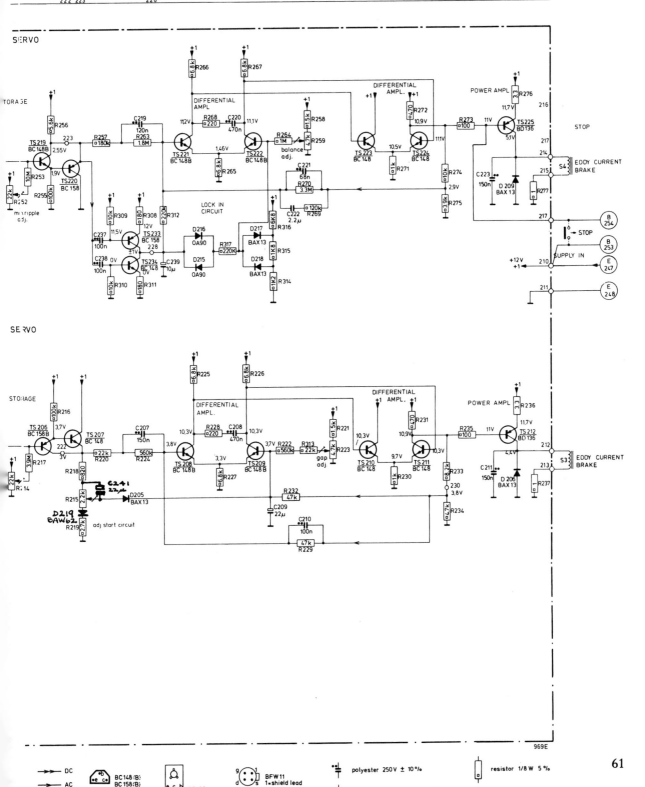

TS219 TS220			TS233 TS234	TS221	D215÷D218	TS222		TS223	TS224		D209 TS225 S4
TS206 D219	TS207	D205		TS208		TS209		TS210	TS211		D206 TS212 S3
	237 238 241	219 207	239		220 208	222 221 209 210				223 211	
252 253 255 256	257	308÷312 263		265÷268	314÷317	264 270 269 258 259		271÷275		276 277
214÷219	220	224		225÷228		222 232 313 229 221 223	230 231	233÷235		236 237
222 223		228							230	210−217	

969E

61

This will change a video head replaying a Ch.1 track to a Ch.2 track and variations in between. The effect is a lateral tracking shift as illustrated in Figure 4.4b.

Capstan servo

The capstan servo, or tape servo, normally has only one function, that is to keep the tape speed constant. It has to do this whilst overcoming varying loads on the tape transport system, such as friction and spool loads within the cassette. In record mode on most machines it runs at a constant speed, but in replay on some video recorders it has to function with the tracking control. Further to this, recorders that have slow motion replay call upon the capstan motor to operate as a stepper motor in short jumps.

Where the capstan servo is used as a tracking control its reference is a stable source and the feedback is from the replayed control track.

We will look at various types of recorder and compare the methods used in drum and capstan servos.

Philips N1500/1501 system

The following section deals with the N1500 servo system in some detail as a grounding for some of the more complex circuitry used by other manufacturers in their machines. Refer to Figures 4.7 and 4.8 and the circuit diagram (Figure 4.9) whilst reading the text.

First, both head and tape servo, in record and play use the same 25 Hz (40 ms period) reference. This reference is a square wave from a divide by two circuit, TS226 and TS227. The input to the divide by two bistable is from a 50 Hz voltage controlled oscillator which is an astable, or multivibrator. It forms part of a phase lock loop called the reference pulse former.

TS228 and TS229 are the voltage controlled 50 Hz oscillator. The output from this oscillator is on the collector of TS229, test point 225. The negative transition of the output square wave is converted to a ramp by TS230, sawtooth generator. Input pulse from field syncs in record and 50 Hz mains reference in replay gate TS231.

TS231 then samples the ramp voltage and the instantaneous level at the time of sample is transferred to C234. This is then the sample and hold technique at circuit level, note the very high impedance circuit values. A DC error voltage is then stored in C234, and an emitter follower, or rather a source follower to be more accurate, supplies the error control voltage to TS228 and TS229. A potentiometer is present to set the free-running frequency of the loop to 50 Hz with TP218 shorted to ground. This is best done in play mode by comparing the waveform on TP225 with mains frequency on a double beam scope and adjusting R293 for minimum drift of one beam with respect to the other.

In record mode the phase lock loop is locked to field syncs which provides the head servo with positional information, feedback is from head PG, K6, via amplifier TS201. Main motor drive is from a synchronous AC motor running at a free speed of about 1510 rev/min at the head drum. It is slowed down to correct speed by an eddy current brake acting upon an aluminium disc under servo control.

A reference from the divide by two circuit, containing field pulse information, is fed to ramp generator TS202. The ramp is sampled from head PG pulses fed via TS203, ramp and sample pulse can be seen on TP220. TS203 is normally 'on' and the base of TS204 grounded, during sample TS203 turns off and the ramp voltage passed via TS204 and TS205 to C206. The sample and hold voltage is then buffered by TS206 and TS207 and is fed to the eddy current brake by DC amplifiers TS208 –TS212.

It can be seen that the tape servo is basically the same. Additional output from TS227 is used to record the control track for replay synchronisation. The differences in the two servos are in the additional 'lock up' drives. The function of these circuits is to hard drive the servos to speed quickly and then allow the phase lock 'sample and hold' to take over.

The replay mode of the servos is identical to record except that the sample pulse on the tape servo is developed from replayed control track. Pulses come in on TP208 after phase shift in the tracking control mounted on another board. By shifting the position of the pulse upon the ramp shown on TP221 the replayed video can be shifted in phase with respect to the mains locked drum. Optimum tracking is then assured.

Let us now consider the additional starting and lock in circuits and their function in each servo. The generated ramp is the reference and is not variable; the ramp is sampled by PG pulses or control track pulses which are variable. At switch on the head will at first run slow so that there will be fewer sample pulses and they will be arriving sequentially late. Looking at the ramp on the oscilloscope will show that the pulses run down the ramp when the head is slow and up the ramp when it is too fast. This produces an AC component across the sample and hold capacitor as in Figure 4.10. During threading some pulses are fed from the threading motor to the drum servo via TP 229 and C240. These additional pulses will provide a high frequency AC component and accelerate the head servo. The slower or faster that the drum rotates with respect to the reference ramp, then the higher frequency the AC component will be. This is supported by the rate that the

pulses run up or down the ramp. That is the faster the pulses run then the higher frequency the AC components will be.

The drum servo has two feedback paths: that via RC232/C209 is positive feedback and that via R229/C210 is negative feedback. If the drum is running very slow, then the pulses will be running down the ramp fast – a high frequency component. For a high frequency component there is a large negative feedback via R229/C210 and gain is low so the servo gradually runs towards lock. As 'lock in' speed is approached the AC component drops in frequency and servo gain increases as positive feedback comes into play. At very low frequencies C209 has a high impedance and the servo gain is very high. In lock there is no large component R232 = R229 and gain stabilises to unity. If there is any tendency to lose lock then the gain becomes high to oppose the change. There is fast start circuit with C241 and D205 and during threading the extra pulses charge C209 rapidly; when lock is achieved D205 is reverse biased as the voltage across C209 is then higher than that set by R215; R215 is set for fastest head 'lock in' time.

Operation of the tape servo utilising the AC component is slightly different, it is slower in action, to prevent tape damage. If the servo is too slow then a step pulse with a high positive transition occurs. This positive step turns on TS234, which turning on, lowers the voltage across C239. TS221 turns off slightly so TS222 turns on, this then turns off TS224 and TS225 reducing the brake current. If on the other hand the servo is too fast a step waveform with a negative transition occurs, turning on TS233 so that the voltage across C239 increases, the brake current will then increase and the servo will slow down. The small AC ripple caused by repetitive sampling will not affect this operation. A time constant C239/R312 prevents rapid overswing whilst diodes D215 to D218 limit the range of control, ensuring stability.

R259 is used to centre the sample pulse on the ramp during record. There is also negative feedback which stabilises the gain frequency characteristic and prevents oscillation of the servo formed by C221, R270, C222 and R269.

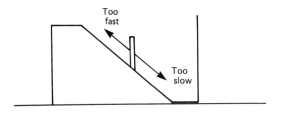

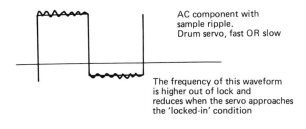

AC component with sample ripple.
Drum servo, fast OR slow

The frequency of this waveform is higher out of lock and reduces when the servo approaches the 'locked-in' condition

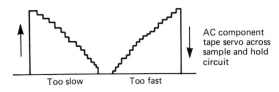

AC component tape servo across sample and hold circuit

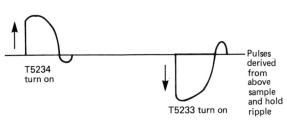

Pulses derived from above sample and hold ripple

Figure 4.10. Waveforms across the sample and hold capacitor

Philips N1700 servo techniques

Figure 4.11 shows the block diagram of the Philips N1700 tape and drum servos. As in previous systems already discussed there is a reference input which is derived from field syncs in the record mode and 50 Hz mains supply frequency in the play mode. Both of these two different references have to be divided by two to 25 Hz, the 40 ms pulse period which is the standard waveform period of most domestic servos.

A slope generator provides the ramp for the drum servo and it is sampled by shaped and suitably delayed pulses derived from K6, which is a PG pick-up head mounted close to the rotating drum. A small permanent magnet mounted beneath the drum induces the PG pulses once every revolution (25 Hz).

The ramp and sample is followed by a storage system to hold the sampled voltage and feed the motor drive amplifiers. The difference in this system from others is the lock-in circuit whose function is to hard drive the drum motor to a speed at which the sample and hold phase lock circuits can take over.

The tape servo also has a slope generator to provide a ramp which is sampled by pulses derived from the tape PG head K4. In the N1700 the capstan revolves very slowly and takes 200 ms to complete one revolution. To ensure proper operation at a period of 40 ms the capstan flywheel has five permanent magnets and each one induces a PG pulse in K4, so these pulses are thus 40 ms apart. As usual the sample and hold circuit is followed by a storage circuit and motor drive amplifier. As the tape drive mechanics have more inertia, and to avoid tape damage by operation that is too rapid, a frequency discriminator circuit is used to run the capstan motor up to speed for the sample and hold phase lock to take over. It can be noted that in the replay mode sample pulses are from the replayed control track and tracking control is within the tape servo operation. This does make response to the tracking control a bit sluggish.

Operation of the lock-in circuit

To understand this system a basic knowledge of simple logic is assumed along with reference to Figures 4.12, 4.13 and 4.14.

The lock-in circuit is supplied with two pulses, reference and sample. The reference is a square wave and so it is shortened in a monostable to single pulse, a positive version and a negative version. First consider the level across C and the output of Q1 and Q2 which is about 7.2 V and fed

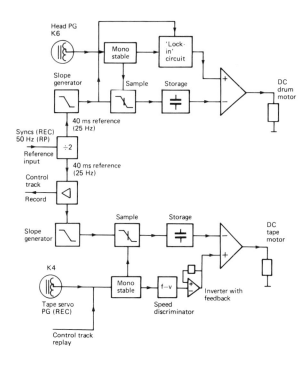

Figure 4.11. Philips N1700 servo block diagram

to the +ve input of the motor drive op-amp. If the level across C rises then the motor will accelerate and vice versa.

Having established that a rise in voltage across C will cause the motor to accelerate we will firstly look at the servo when it is running slow. In the locked or normal running mode flip-flop 1 is set by the reference pulse and then reset by the sample pulse, inverted for the process. The square wave output from flip-flop 1 is compared with an inverted reference pulse in a NOR gate (see Figure 4.13). Provided that the sequence between the reference and the sample is Set–Reset, then the Q output of flip-flop 1 will always be 'high' when the inverted reference is 'low'; and the reference will be 'high' when the Q output of flip-flop 1 is low. The NOR gate output will thus remain low as either input is high at any time. However when the servo runs too slow sample pulses will arrive later and later and there will be a time when the Set–Reset sequence is lost. Note that the reference pulse to flip-flop 1 is connected to the 'clock' input and so it can set *and* reset the flip-flop. When the condition is reached that a sample arrives too late, then the reference will set, and subsequently reset, flip-flop 1. It can be seen from the timing diagram

64

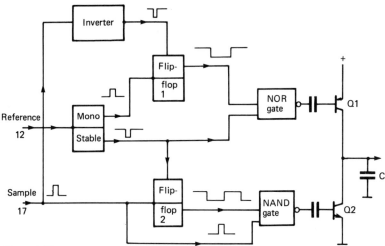

Figure 4.12.

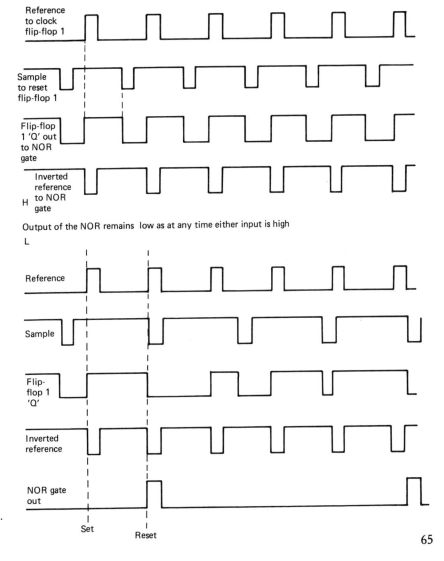

Reference
to clock
flip-flop 1

Sample
to reset
flip-flop 1

Flip-flop
1 'Q' out
to NOR
gate

Inverted
reference
to NOR
gate

H

Output of the NOR remains low as at any time either input is high

L

Reference

Sample

Flip-
flop 1
'Q'

Inverted
reference

NOR gate
out

Figure 4.13.

Set

Reset

65

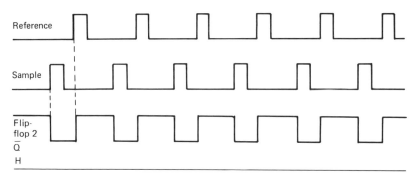

The output of the NAND gate is 'high' as neither the sample nor flip-flop 2 $\bar{Q}$ are high at the same time

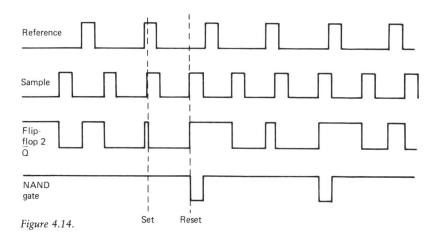

Figure 4.14.

that under these conditions the Q output of flip-flop 1 will be low at the same time an inverted reference pulse is low. The NOR gate then has two low inputs and its output will go high. This provides a series of positive pulses over a relatively longer time period to turn on Q1 and increase the level across C.

If the servo is only running slightly slow, only a few pulses occur. If on the other hand the servo is running very slow then many pulses will occur. This produces the relationship across C, that the more out of lock the servo is then the greater is the increase of voltage across C to correct for the error. If upon switch on, the servo is running very slow then the level across C is very high; as the motor accelerates and the difference in the 'locked-in speed' and 'running-speed' reduces, the level across C reduces. When the 'running speed' equals the 'locked'in speed' the 'lock-in' circuit drops out of operation as the NOR gate output remains low.

Flip-flop 2 operates in the same way but has inverse pulses and a NAND gate. Flip-flop 2 is set

by sample pulses and reset by reference pulses and a square wave is produced on its Q output. Again, provided the Set–Reset sequence is maintained then any one input to the NAND will always be low at any time (see Figure 4.14) and its output will always be high. If the motor or servo is for some reason running too fast flip-flop 2 will be set and then reset by two sequential sample pulses, causing two high inputs to the NAND gate so that its output goes low.

A pulse is then created to turn on Q2 and the voltage across C lowers. This reduction of voltage across C reduces the motor speed and so the servo slows back into lock. Again once the locked-in condition is reached the lock-in circuit drops out and the output of the NAND gate remains high.

Tape servo frequency discriminator

This is an introduction to a technique which will become the standard method of hard driving the servo to the correct speed for the phase lock sample and hold to take over. This is the dual loop system and we shall be looking at this in later video recorders.

The discriminator consists of two monostables, a ramp generator and a switch with a storage capacitor at the output, not too dissimilar to a sample and hold circuit. PG pulses from the capstan pick-up head, K4, are fed via a monostable to the frequency discriminator circuit. In the discriminator Mono 1 is triggered for a short period, when it resets it triggers Mono 2. Mono 2 starts a ramp from high towards low, and the rate is constant. After 40 ms a second PG pulse arrives, samples the ramp voltage at that time and then resets the ramp back high. Therefore if the servo motor speed is constant then the voltage across storage capacitor C will remain constant. If the motor runs slow then PG pulses will start to arrive later in time, because the ramp slope is constant

later arrivals will sample the ramp at lower and lower voltages, therefore the voltage across C will fall. Conversely, if the speed of the motor increases then PG pulses will start to arrive earlier; this will not give the ramp much time to decrease and it will be sampled and reset at a higher voltage, consequently the voltage across C will rise. Figure 4.15 shows the block diagram and Figure 4.16 the timing diagram.

The frequency discriminator is then a converter from frequency to voltage shown as $(f - v)$. The output of C, as it stands, is positive. That is to say an increase in speed causes an increase in voltage, but this is the wrong polarity – the motor has to be slowed down. This is achieved by taking the voltage across C into an op-amp inverter and providing frequency dependent feedback for stability. The output of the op-amp then falls if the motor speed increases and this then reduces the

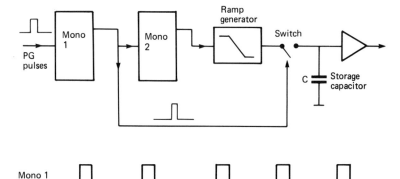

Figure 4.15. Frequency (speed) discriminator block diagram

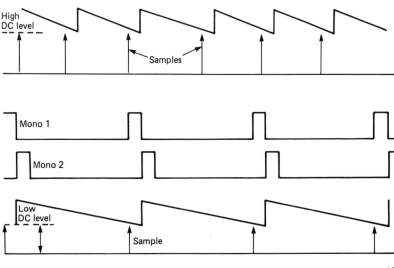

Figure 4.16.

output of the motor drive amplifier thus reducing the motor voltage and correcting for the increase in speed.

At 'switch on' the voltage across C will be low so the motor drive voltage will be high; as the motor approaches lock-in speed the drive voltage falls and the sample and hold circuit takes over. Note that the frequency discriminator still stays 'on line' and works with the phase lock circuits giving the dual loop servo technique.

HR3300/3330/3320

The first VHS videorecorders to come on to the market were manufactured by JVC although they came in various disguises. The Ferguson Videostar is one, and it had a performance slightly less than its JVC counterpart by the exclusion of an equalisation filter driving the RF modulator. Major rental outlets also marketed the basic VHS recorder.

The servos of the basic VHS videorecorder are shown in the usual block diagram in Figure 4.17.

The capstan servo is designed for constant speed and uses a phase locked servo technique to achieve this. A 'tuning fork' oscillator is used as a reference and this reference is the sample of a ramp.

A PG pulse from the capstan flywheel is squared and turned into a ramp, which is sampled by the reference oscillator and the resultant DC error is used to control the capstan DC motor as previously detailed. This, then, is a technique where a stable reference is used to 'pace' a servo and maintain constant speed. Control of the motor is by a discriminator circuit, not to be confused with a frequency discriminator – its function is described in a later paragraph.

Very little need be said about the drum servo, which is identical to the standard servo as previously described. Mono 1 and Mono 2 are very short monostables and are used to delay the PG pulses slightly so that the mechanical tolerances of the flywheel and magnets, upper head assembly etc, can be adjusted out and the head switching points defined accurately. The resultant square wave from the flip-flop is a very important signal as it is used not only to switch video heads on replay but also to determine phase switching of the

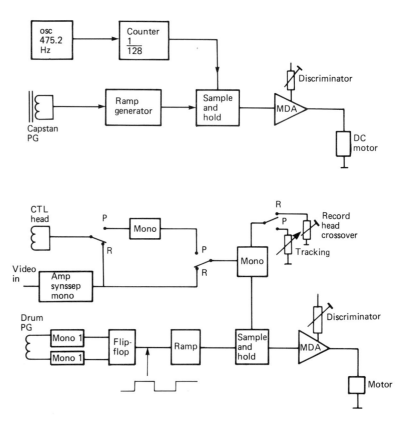

Figure 4.17. Block diagrams for the HR3300/3330/3320 – top, capstan servo; bottom, drum servo

colour circuits (see Chapter 5). This waveform is also utilised for the ramp and it is sampled by frame pulses in record and replayed control track pulses in replay.

A more notable feature is the discriminator part of the motor drive amplifier as shown in Figure 4.18. The discriminator IC and the motor drive transistors are lumped together as one amplifier with a gain of about unity. Consider that the servo is running, and for some reason there is an increase in friction either in the tape path or a tightening of tape around the drum. In this case the current in the drive motor will rise as it is loaded by the friction. An increase in motor

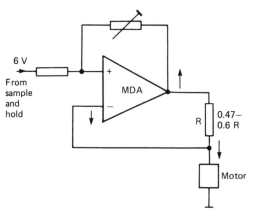

Figure 4.18. Discriminator block diagram

current causes a voltage drop across R, shown by the small arrow. The fall in voltage across the motor is fed into the inverting input of the MDA. With a gain of unity the voltage change arrives at the output of the MDA but inverted and at the same value as the fall in voltage. This cancels out the reduction in motor voltage caused by loading of the motor and has the effect of providing the motor with the extra current it needs if required without having any changes in speed, which of course would then cause the whole servo loop to change rate.

Another part of the servo loop is the 'non-linear' circuit (Figure 4.19), which has many forms in the VHS range of machines but its function is always the same. It is found only on tape drive servos and functions in the play mode only. It is rather like two back to back zener diodes of a volt each, and it can also be found in the form of an NPN, PNP transistor network. In replay mode the servo is being paced by replayed control track pulses which were recorded in the record mode. The tape speed as it travels through the recorder is subject

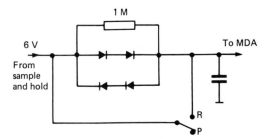

Figure 4.19. Non-linear circuit

to minor fluctuations so control track pulses can be recorded with slight timing errors. Also they can be replayed with slight timing errors. As this is the case the sample and hold output is varying continuously around the 6 V DC level; bearing in mind that the tape servo is slow in its response, then it would not be a good idea to allow such minor variations to affect the servo loop. The non-linear circuit allows the sample and hold to vary about 1 V up or down with respect to the standing DC voltage without changing the motor speed. This is due to the forward voltage of the diodes being about 1.2 V, so the variations within this range are eliminated by not being allowed through the servo.

In the Philips N1500 VCR each motor was run at a speed slightly higher than the 'locked-in' speed and slowed down, while in the N1700 there was a specific 'lock-in' circuit. In the JVC VHS recorders there is a limiter circuit to define the general speed of the motor and provide a range within which the sample and hold phase lock can work. It is detailed in Figure 4.20.

The voltage across the drum and capstan motors is 6 V, a reference of 6 V is defined by two 10K resistors and a 560R resistor between them. Diodes D1 and D2 will each have a forward voltage of 0.6 V; D1 will not allow the servo voltage to rise higher than 6.6 V and D2 will not

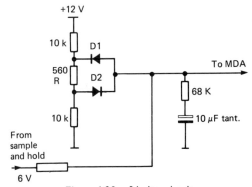

Figure 4.20. Limiter circuit

allow the voltage to fall below 5.4 V. At switch on the motor is driven at 6.6 V until the servo locks up and the drive voltage stabilises to 6 V and the phase lock system will then operate between these limits.

The limiter circuit along with the gain settings of the discriminator will provide for minimum lock-in time and prevent the servo from 'hunting' outside the limits set.

HR4100 portable recorder

Completely different approaches are adopted for the portable video recorder and much tighter control is used, especially on the drum servo. This will be subject to gyroscopic effects, that is to say that the recorder will be moved around whilst in use and it would not be advisable to allow the external movement to affect the drum rotation. The HR4100 and its equivalents have a sample and hold arrangement similar to that used in previously discussed machines, the difference being in a frequency generator (FG) technique. The frequency generator is a method of holding the speed of the motor at a value within which the sample and hold phase lock system can work. It is used as an active system, similar but not identical to the 'lock-in' circuit of the N1700, since the N1700 'lock-in' circuit drops out of action when the correct speed is reached, in the FG principle it does not, but remains continuously active. It is

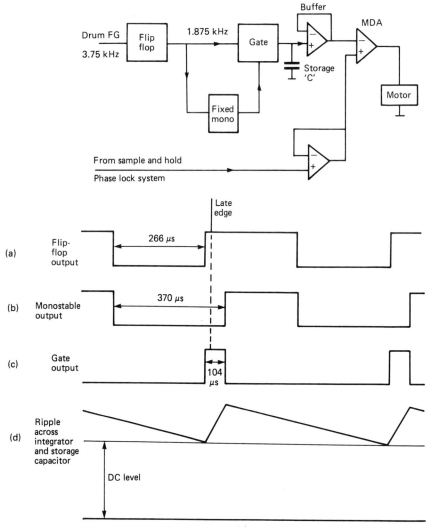

Figure 4.21.   HR4100 FG block diagram (top) and timing diagram (bottom)

additional to phase control and provides tighter control.

Built into the drum assembly is a frequency generator formed by many teeth passing a coil. In the HR4100 the FG is 3.75 kHz (3750 Hz) and the output is amplified and divided by two in a flip-flop to obtain a square wave which is symmetrical. The rising edges of the square wave triggers a monostable the timing of which (b) is set to a specific value, 370 μs. The timing of the output of the flip-flop is 266 μs (1875 Hz), (a) and the two values are compared in a gating circuit. An output

pulse from the gating circuit is then the difference of the two inputs (c). The pulse is integrated and stored resulting in a DC level which is dependent on the length of the pulse (d). If the motor runs slow then the timing period from the flip-flop will increase as shown in (a); the mono timing is fixed therefore the output pulse (f) from the gate will shorten with a resulting fall in DC voltage across the storage capacitor. An op-amp after the storage capacitor is used as a buffer and it is fed to an inverting input of the MDA op-amp. This means that a fall in DC voltage across the storage capaci-

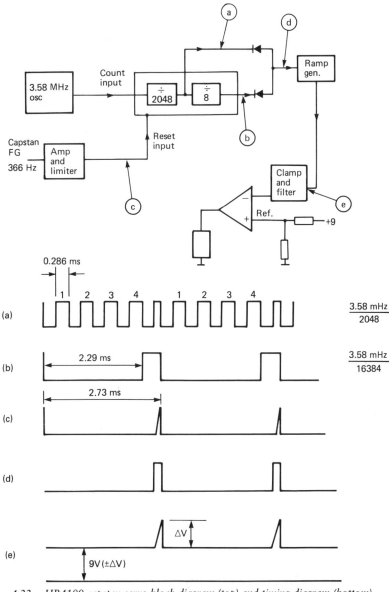

Figure 4.22.   HR4100 capstan servo block diagram (top) and timing diagram (bottom)

71

tor will increase the motor drive voltage to compensate. An increase in motor speed will give opposite results.

The FG section of the servo may be considered as coarse control and the phase lock side as fine control, the combined result being a much more stable and tighter controlled servo compared to previous methods.

An entirely different approach is used in the capstan servo, a technique not commonly employed (Figure 4.22).

As pointed out already, any servo must have two inputs, a reference and a feedback. In this particular servo the reference is a 3.58 MHz oscillator counted down to approximately 1748 Hz by a division of 2048 (a). The capstan FG output is about 366 Hz; it is shaped and used to reset the counter chain. Two outputs are obtained from the counter, 3.58 MHz divided by 2048 and the end of the chain (b) which is a further count of 8. The two outputs are combined in an AND gate thus the resultant pulse (d) is a combination of (b) and the fifth pulse of (a). Control of the servo is by the pulse length of the pulse (d) and that is determined by the reset pulse (c) and when it arrives to reset (b). The pulse (d) charges a capacitor which is clamped and filtered to control the motor speed through the inverting input of the motor drive op-amp. Operation is then, if the motor is slow, the reset pulse (c) is late and thus (d) is longer and the level of voltage (e) is higher, which then slows the motor through the op-amp inverter.

This is an example of a pulse counting technique as applied to speed control as no phase control is required.

Drum speed correction

Recent machines are fitted with a *picture search* feature, in which the tape can be replayed at about ten times normal speed both forward (cue) and backward (review). Except in some V2000 format machines this introduces tracking errors, and synthesised vertical sync pulses must be inserted into the video signal (See Chapter 7) to provide a stable reference for the TV field timebase. More important is the fact that the head to tape speed alters; in cue mode the tape speed is accelerated. Consider a video head scanning a tape track in cue mode. The tape, travelling in the same direction as the head at higher-than-normal speed, may be

regarded as 'running away' from it. To catch up, as it were, the video head would have to increase speed – at normal head velocity the head to tape speed would be reduced, with a corresponding fall in line sync frequency of about 5%. Unless this is compensated for by a 5% increase in head drum speed the result will be loss of horizontal hold on the display TV. Conversely, in review mode the tape changes direction and passes 'backwards' round the video head drum, moving towards the advancing video head. The result is an increase in head to tape speed of about 6%, necessitating a 6% reduction in head drum speed to maintain correct line sync frequency.

In the first machines to feature picture search the head drum was not servo-locked in cue and review modes – the motor was merely switched between two appropriate preset speeds. Later models used more elaborate circuits in which replayed line syncs are measured in a frequency-to-voltage convertor (*frequency discriminator*) whose output drives the drum motor to compensate.

There are two methods of arranging the fast tape transport (approx. ×10 speed, forward or reverse) required for search modes; Panasonic favour running the capstan servo at 9× play speed, giving the advantage of locked mistracking-bars during search, whereas JVC and others prefer to run the reel motor at about 10× normal speed by comparing its speed with that of replayed control track pulses. Reel motor speed is governed by a frequency-to-voltage convertor: if the motor speed falls the control track pulse rate falls, and since the f-v converter is arranged in inverting mode its output voltage rises to restore correct motor speed. Details are given later in the section 'Further uses of frequency discriminators'.

There is an unfortunate side-effect of drum speed correction in picture search modes, however. We have seen the necessity to inject synthesised vertical sync pulses for TV field stability; they are derived from the edges of the drum flip-flop squarewave via very short time delay monostables, the period of one of which can be adjusted to eliminate picture bounce in still-frame mode. As a result of the required 5% increase in drum speed during cue mode it follows that the field sync pulse frequency will also increase by 5%. Due to the direct-sync system employed in most TV sets this can be accommodated, whereas the reduction in field frequency by 6% which occurs in review mode is quite a different matter – most TVs cannot cope with it, and field roll is the result.

Dual loop servos

All modern servos have two loops, one for phase control and one for speed control. Earlier in this chapter we studied basic servos and phase control, where motor speed was controlled by a fixed voltage divider called a limiter (Figure 4.20). Then we progressed to the example of the JVC HR4100 videorecorder with its additional *FG* (frequency-generator) loop for speed control. As design improved the original limiter technique was abandoned in favour of a dual-loop servo system as shown in Figure 4.23. It has a similar effect to that of the tape servo frequency discriminator of our Figure 4.11. The pulse position control at the top right-hand side of Figure 4.23 acts to set the sample pulse position upon the ramp in the sample and hold circuits; a more detailed description is given later in the section dealing with further uses of frequency discriminators.

locked loop, taking in the sample/hold section, motor drive amplifier, motor and PG generator.

In order for this phase control loop to lock, the motor must be running at a speed very near correct. The situation is akin to that in the flywheel line sync circuit of a TV (which contains a similar phase lock loop) where the frequency of the set's line oscillator must be very close to incoming sync rate to achieve phase-lock. In the same way as line-hold adjustment on the TV will shift the image *relative to the raster*, so varying the phase of the VCR drum servo loop will alter the relative position of the video heads at the arrival time of an incoming field sync pulse. In Figure 4.5, for instance, the position of Ch.1 video head can be set so that it is at any required point (within limits) on the circumference of the drum when it records the field sync pulse. Normally it is set for about 7 TV lines after the head enters onto the tape by a variable delay pot often called 'Record Switch Phase'.

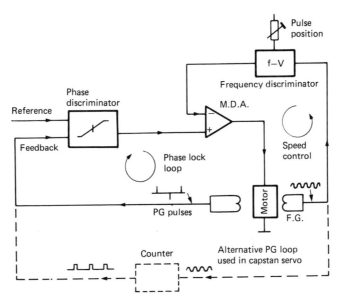

Figure 4.23. Dual loop servo

Figure 4.23 contains all the elements of a servo system which became very widely adopted for use in both drum and capstan servo systems. The main features are the input reference, input/feedback control loop, and speed control loop. The reference input is a fixed frequency (i.e. *clock*) signal to which the servo locks, while the feedback loop contains 'feedback from the rotating device'. The whole forms an electro-mechanical phase

In a dual loop servo the two loops work together, with the FG (Frequency Generator) loop setting the speed to a level within which the PG (Pulse Generator) phase control loop can lock. Figure 4.24 shows the relative gain of the two loops and illustrates their mutual operation. When the motor speed is low the FG loop gain is high to give fast acceleration; as correct speed is approached the FG loop gain falls while PG loop gain increases to give very fast and stable phase-lock, at which PG loop gain is at maximum. Note that if motor speed becomes excessive the FG loop gain once more increases, but this time in the correct sense to brake the motor.

Motor behaviour is monitored in two ways: as a single pulse per revolution for the PG loop; and as a continuous 'tone' signal whose frequency is proportional to speed for the FG loop. Regardless of whether these signals are generated within the motor or from driven members they represent 'feedback from the rotating device'.

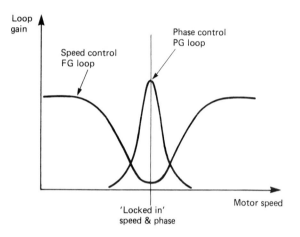

Figure 4.24. PG and FG interaction

The need for 'positional' information is plain in the case of the drum servo, whereas the angular position of the capstan shaft at any given instant is not important. For capstan control, then, it is common practice to derive 'PG' pulses by counting down the FG signal to a frequency commensurate with the incoming reference – see also Figure 7.22, p.126.

	Reference	Feedback	Speed
Drum			
Record:	Vsync/2	Drum PG 25 Hz	Drum FG 1600 Hz
Playback:	Xtal 25 Hz	Drum PG 25 Hz	Drum FG 1600 Hz
Capstan			
Record:	Xtal 21 Hz	Cap FG 21 Hz	Cap Motor FG 870 Hz
Playback:	Xtal 25 Hz	CTL PB 25 Hz	Cap Motor FG 870 Hz

$$21\,HZ = \frac{CAPSTAN\ FG\ (252.5\ Hz)}{12}$$

Crystal Clock Frequency – 4.433619 Mhz

Table 4.1

Reference signals

During record the reference signal for the drum servo *must* be incoming field syncs to ensure correct field/track registration. In playback the reference can be based on off-tape control track pulses or a crystal clock via a count down circuit; in the latter case control track pulses will be applied to the capstan servo. See Table 4.1 for details of reference and feedback sources for both servos.

Introduction to digital servos

As we have seen, the conventional analogue servo is based upon the principle of sampling a voltage ramp to provide a DC voltage proportional to the time-lapse between the arrivals of the ramp-triggering pulse and the sampling pulse. Such ramp-and-sample circuits may be regarded as either a phase discriminator (timing variation corresponds to phase shift) or as an APC (Automatic Phase Control) loop. The output from this type of circuit is a variable DC voltage carrying a superimposed ripple waveform due to the instantaneous sample pulses. It is passed through a DC amplifier for presentation to the drive motor as an error voltage.

Digital servo design is based upon an entirely different principle, and while a DC error voltage is produced, it is derived in a very different way. Instead of a ramp and sample circuit a digital counter is used to measure time delays or variations. The idea is for the reference pulse to set the counter going from zero (corresponding to ramp initiation) and then to stop its count some short time later by means of the sample pulse. The accumulated count at this time is indicative of timing delay: if the sample pulse arrives early a low count will be registered, whereas if it comes late a correspondingly high count will be accumulated. The use of additional circuits permits the count length to reset a bistable which was set at the start of the count. Bistable output is a squarewave, and if *reset* is arranged to take place half-way through the period between consecutive 'sets' this squarewave has equal (1:1) mark-space (m/s) ratio. If the count time subsequently varies then the m/s ratio of the squarewave varies accordingly.

An integration filter whose input is a square-wave with equal m/s ratio produces at its output a DC level of half the peak input voltage, i.e. if its input is switched between ground and a 12 V rail,

6 V would appear as an output. Any variation in incoming m/s ratio would reflect in the DC output voltage so that for instance a 3:1 m/s ratio would render 9 V DC and a 1.3 m/s ratio 3 V DC. In this way a variable control voltage can be derived from a digital servo and fed to a motor as an error voltage in the same manner as for analogue servos. A simple example of a digital servo is furnished by the Toshiba system described next.

Toshiba V5470 servo

In the first Toshiba digital servo the place of the 'ramp', then, is taken by a stored 0–10 bit count. The amount of count in the store is governed by the two inputs to the servo, reference and sample. The stored count is read out as a high frequency squarewave with variable m/s ratio. This m/s ratio is varied at the sampling rate of 25 Hz although the square wave itself has a frequency of 1.45 KHz.

Referring to Figure 4.25, the reference input clears the memory and resets the 15 bit counter which then starts to count clock pulses at a rate of 746.56 KHz. After about 1.3 ms a sample pulse arrives and stops the count, which is committed to the 10-bit memory. Here it remains until the next reference pulse some 40 ms later. During this period the contents of the memory are loaded into one half of a comparator. Meanwhile a second 10-bit counter has been counting 'double speed' clock pulses at 1.493 MHz, continuously filling up and resetting. Each time it resets an output pulse is passed to the Set/Reset (SR) bistable to set its output high. The running count in the second counter is continuously available to (and monitored by) the second half of the comparator. When the counts in each side of the comparator match, a reset pulse is generated and made to reset the bistable output low.

The proportionate length of time that the SR bistable output is 'high' compared to its 'low' period is dependent on the count stored in memory. Therefore the longer the time lapse between reference and sample pulses the greater will be the stored count and the longer the 'high' period of the bistable. As we have seen, the next stage is an integrator, and its output voltage rises in these circumstances to speed up the motor, counteracting whatever influence slowed it down to cause the sample pulse to arrive late in the first place.

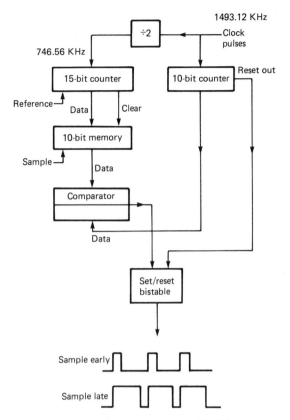

Figure 4.25. Toshiba V5470 servo block diagram

This servo has inbuilt limiters; if the motor is running fast out of lock the bistable output remains low, and if the motor is running slow the output becomes permanently high to maintain full drive to the motor.

There is one problem inherent in this design of digital servo: its 'lock-in' time is slightly longer than that of an analogue servo. Because the memory is refreshed at 40 ms periods (25 Hz rate) and the bistable rate is 1.46 KHz, the comparator output is about 60 times the memory refresh rate; to put it another way the contents of the memory between reference pulses are used 60 times. The input clock rate to the 10-bit counter is 1.493120 MHz. A 10-bit counter is a 2^{10} divider, or divide by 1024, and hence its output reset rate is 1.458 KHz.

Digital versus Analogue

Why change to new digital techniques when analogue systems have been tried, tested and developed to a high level? There are several advantages inherent in digital servo systems even though they do the same basic job as their analogue counterparts. Capacitors used for pulse timing and ramp forming can be eliminated and replaced by stable and reliable electronic counters. LSI (Large Scale Integration) techniques permit the incorporation of many counters – and associated peripheral logic functions – in a single chip. This opens the way to accommodating both capstan and drum servos on a single IC allowing for separate or shared references which can then be switched or selected with ease; hence increased versatility and a potential for more user features. Also significant is the reduction of preset controls; reduced component count; and the high intrinsic timing accuracy associated with digital designs.

Second generation digital system

Let us now turn to a more highly developed digital servo which uses similar principles to the Toshiba design already described, but offers more refinements. Its configuration is outlined in Figure 4.26. During record the drum servo digital comparator locks the drum PG feedback pulses to incoming field syncs, suitably divided by 2. As in an analogue servo, speed control is achieved by a frequency discriminator, though the technique used here is different. The digital comparator output operates on an SR bistable whose PWM (Pulse Width Modulated) output depends, as before, on the timing of its set and reset pulses. A passive low-pass filter then smooths (integrates) the PWM output to a varying DC voltage for application to the motor drive amplifier.

In practice a single IC caters for both capstan and drum servo operational circuits. Many of their

functions are common or interrelated, so compared to previous models the two servos are less easy to divorce. Table 4.1 identifies the sources of reference, feedback and clock signals. In record the drum servo locks the drum PG of 25 Hz to incoming field syncs divided by two to 25 Hz, whilst the FG tone (1600 Hz) controls the drum speed via the now-familiar frequency discriminator. In replay the drum PG is locked to a 4.43 MHz crystal oscillator counted down to 25 Hz; the FG loop operates as in record mode.

The capstan flywheel FG generates (at normal speed) a 252.5 Hz tone which is counted down to 21 Hz, enabling it to be locked to a 21 Hz reference derived by counting down the 4.43 MHz crystal clock. A second capstan FG is incorporated in the motor to generate an 870 Hz tone for use in the speed-controlling frequency discriminator. In playback the capstan servo works at 25 Hz to match the off-tape control track pulses; these form the feedback source to be locked to a 25 Hz reference counted down from the 4.43 MHz crystal clock.

During recording there need be no specific relationship between drum and capstan servos, and here the capstan servo simply controls the linear tape speed to 23.39 mm/sec in accordance with the VHS format specification. During playback, however, both drum and capstan servos are tied to the common crystal oscillator reference source; it is compared with drum PG pulses in the drum servo, and with off-tape control track pulses in the capstan servo. The common reference signal provides the 'phasing' link between capstan and drum shafts which is necessary for consistent tracking with minimal error – the drum runs at constant speed and phase while the capstan corrects for record/replay jitter characteristics.

Having examined the inputs to these servos, and seen that both phase and speed loops are

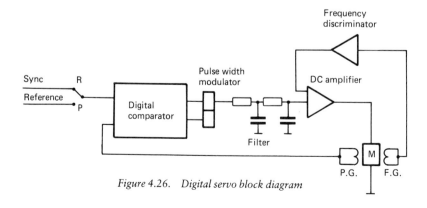

Figure 4.26. Digital servo block diagram

incorporated, we have set a scene which could equally apply to an analogue servo system. Time now to investigate the heart of this new servo arrangement in terms of counters, memories and latches.

Digital principles

The basis of the JVC HRD120 servo is a 10 bit counter which counts from 0000000000 to 1111111111 in a binary count of 2^{10} which is: $2^9 + 2^8 + 2^7 + 2^6 + 2^5 + 2^4 + 2^3 + 2^2 + 2^1 + 2^0$.

Multiplying this out we get: $512 + 256 + 128 + 64 + 32 + 16 + 8 + 4 + 2 + 1 = 1023$. As the first step in the counter is 0000000000 and the last is 1111111111 then to complete the count back to 0000000000 a further 1 step is required and so the full count is 1024. Let us now assume that the total count from ten zeros to ten ones is the equivalent of the ramp in an analogue servo, as depicted in Figure 4.27. In the analogue system the ramp is sampled at its halfway point, and halfway through a 10 bit count comes binary word 0000000001 – here the MSB (Most Significant Bit) is on the right.

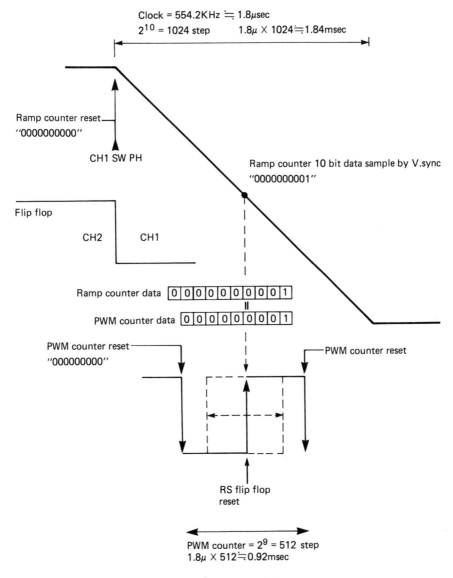

Figure 4.27. Digital ramp

In the full block diagram of Figure 4.29 there are four elements: a 10-bit ramp counter, a 10-bit latch memory, an equality detector and a 9-bit pulse width modulator counter. Before getting to grips with these in detail an overall view is required. Referring back to Figure 4.27 the ramp counter is reset to 0 by a Ch.1 switch phase edge pulse at the instant that Ch.1 head is switched on, whereupon the counter commences its count. When it is about half full a sample pulse from vertical sync transfers the ramp count into a 10-bit latch memory. Meanwhile the second (PWM) counter has filled up, reset (setting the bistable as it does so) and started to count up again. When its count data matches the ramp counter data (held in the latch memory) an equality detector resets the bistable. Plainly, any timing differentials will alter the count word held in memory and modify the bistable's m/s ratio.

We can now go on to look at the system in some detail in Figure 4.29 and see how the digital techniques mirror the analogue servo system while conferring the advantage of better stability. Starting in the top left-hand corner, the 4.43 MHz crystal clock is counted down in a $\div$ 8 circuit to 554.2 KHz which as a period of 1.8 μs. This 1.8 μs clock pulse is fed to the 10-bit ramp counter via switch SW1. The full 10-bit count of 1024 thus requires a period of 1.8 × 1024 = 1.843 ms, and this is the duration of the pseudo ramp – see also Figure 4.27. So – it takes 1.84 ms to completely load the 10-bit ramp counter with 1's; the halfway point in this count occurs when bit 10 is set to 1 and the other nine bits are all at 0 – corresponding to a count of 512. Note that (like all the other counter elements) bit 10 is basically a bistable ($\div$2), and it having taken 512 counts to *set* the tenth bit it is plain that a further 512 counts are requires to *reset* bit 10. Hence the halfway count to a bit status of 0000000001.

We have seen in this chapter that in a drum servo during record mode an *analogue* ramp would be initiated by the flip-flop transition between Ch.1 and Ch.2 switch phases, then sampled at its halfway point by a pulse derived from incoming field syncs. In a digital servo the ramp counter is reset to all zeros by a Ch.1 switch pulse, then 'sampled' after a count of 512 by a vertical sync pulse. This sampling action is achieved by a *sample pulse generator* driving the 10 pole latch switch to transfer the contents of the ramp counter at that instant into store in the 10-bit latch memory. The latching action is repeated at 40 ms intervals, with Ch.1 switch

pulses regularly resetting the ramp counter to start from zero, only to have its contents 'dumped' into the latch memory within 1.84 ms by the appearance of the vertical sync pulse. The normal timing for an operational and locked servo is 0.92 ms, corresponding to a halfway count of 512 from reset; this of course gives a symmetrical ± range of correction. The ramp count is 'frozen' in the latch memory in the 40 ms period between refreshes. This action is similar to that which takes place in the storage capacitor of the analogue sample and hold system, where refreshing/updating also takes place at 40 ms intervals. To sum up this section of our digital system in Figure 4.29, then, the ramp length is 1.84 ms; it is normally sampled halfway at 0.92 ms to transfer a binary word of 0000000001 to the latch memory.

Meanwhile the PWM counter at the top of Figure 4.29 is counting continuously. This 9 bit counter has a capacity of only 512 (111111111) so that when fed with the same 1.8 μs clock pulses it will fill up in 1.8 × 512 = 0.92 ms. After 0.46 ms from start it will be half full with a binary status of (0)000000001 which matches the word stored in the latch memory. The 10 bit equality detector responds to the match between the bit pattern in its two registers by producing an output pulse to reset the PWM bistable. Note that the input 554.2 KHz to the 9 bit PWM counter contributes the 'first' zero to the binary word match in the equality detector.

To sum up the operation of the PWM counter: it sets the PWM bistable each time it 'overflows' and resets to zero. Halfway through its count there occurs a match in the equality detector which resets the bistable. The PWM continues its count until auto-reset when it again sets the bistable. Therefore providing that the latch memory holds a binary count of 0000000001 (corresponding to the halfway ramp count) the equality detector will reset the bistable halfway through the PWM count, resulting in an equal (1:1) m/s ratio in the bistable output.

Operation

The corrective action of this servo is illustrated in Figure 4.28. If the drum motor tends to slow down, the ramp counter reset is late, and as a result little count is accumulated – it is less than half full by the time an incoming vertical sync pulse arrives to transfer its count to the latch memory. This low binary count duly appears on the lower register of the equality detector, and the

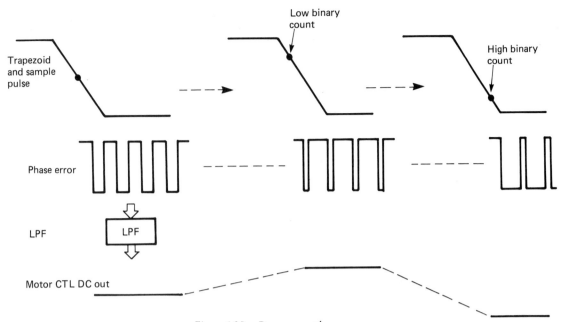

Figure 4.28. Drum servo phase error

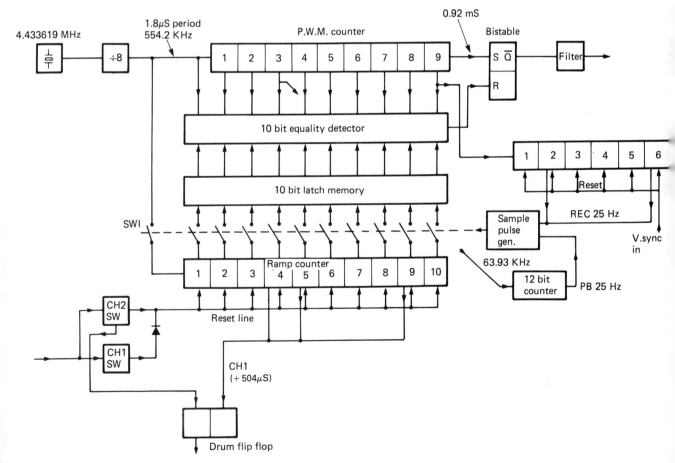

Figure 4.29. Digital servo principle

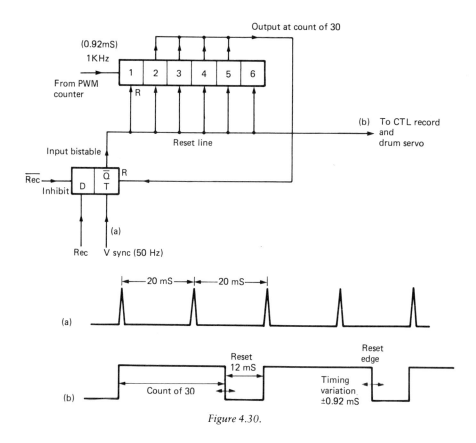

Output at count of 30

(0.92mS)
1KHz

From PWM
counter

| 1 | 2 | 3 | 4 | 5 | 6 |

R

(b) To CTL record
and
drum servo

Reset line

Input bistable

Rec̄

Inhibit

D | Q̄ T | R

(a)

Rec   V sync (50 Hz)

|← 20 mS →|← 20 mS →|

(a)

Reset
edge

Reset
12 mS

(b)   Count of 30

Timing
variation
±0.92 mS

Figure 4.30.

PWM counter (after resetting itself and setting the bistable) does not count far before a match is made within the equality detector to quickly reset the bistable once more. As a result the bistable output does not dwell long in its 'low' state – most of its 0.92 ms period is spent in a high state. This high m/s ratio integrates out to a higher DC level than would a 1:1 m/s ratio, so that motor drive is boosted. Figure 4.28 shows the comparison between the PWM/LP filter behaviour and an analogue ramp and sampling system.

Further advantages of digital techniques

Earlier in this chapter we discussed the use of a monostable device to effectively divide the field sync repetition frequency by two. The monostable is triggered by a field sync pulse and as its period is set to 28 ms the next field pulse to arrive – 20 ms later – is ignored. At 28 ms the monostable once again becomes armed, in time for the third field sync pulse to trigger it again, only to 'flop' after a

further 28 ms. Thus is built up at the monostable output a squarewave with m/s ratio of 28/12 ms and a total period of 40 ms – from incoming syncs at 20 ms intervals. The pulse rate is divided by two, but the output is not symmetrical.

A similar ÷2 function can be achieved digitally. At first sight the method appears more complex, but in practice large numbers of counters and gates are easy to fabricate in a large scale IC, and this approach does not call for a timing capacitor. Figure 4.30 illustrates the principles involved; the circuit block is also shown (but with less detail) on the right-hand side of Figure 4.29. An inhibit signal to the input bistable is removed during record, enabling the first input field sync pulse to set the bistable, which in turn outputs a reset signal to a 6 bit counter being clocked by a 1 KHz signal. In fact the clock signal is the 0.92 ms output from the PWM counter, so it is slightly above 1 KHz in frequency.

It is arranged by the logic output of the counter chain that at a count of 30 the input bistable is reset; it remains in that condition until the arrival of the next field sync pulse. The set period of the bistable is therefore 30 × 0.92 = 27.6 ms, a figure

80

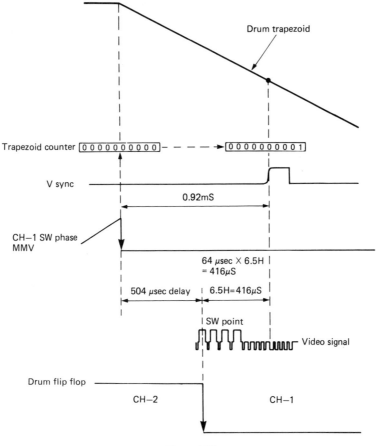

Figure 4.31.

very close to the 28 ms period used for the bistable circuits of earlier designs. Oscilloscope examination of the output waveform from the digital ÷2 circuit reveals a degree of timing jitter on the reset edge, manifest as a double-edge shimmer — it is due to the fact that the 1 KHz clock and the incoming field sync are not synchronised, allowing the correlation between reset edge and field sync to vary by 0.92 ms.

Another timing function in analogue servo systems is the record switch phase adjustment. You will remember that its purpose is to set the drum flip-flop (switch phase signal) edge to occur 6.5 lines before field sync pulse during record. It can only be correctly set when the Ch.1 and Ch.2 switch phase adjustments (replay mode) have been set up. In a digital servo system it is possible to eliminate the record switch phase pre-set altogether by introducing a short delay in the flip-flop signal.

We have already seen that the output from Ch.1 switch phase monostable resets the ramp counter and then halfway through the count it is 'sampled' by field sync, with a time lapse of 0.92 ms between these two events, Figure 4.31. If a suitable delay is introduced between the Ch.1 switch phase monostable and the RF switching flip-flop, the Ch.1 flip-flop edge can be made to occur 6.5 lines before field sync. The time taken for 6.5 TV scanning lines is $6.5 \times 64 = 416 \,\mu$s, and the time lapse between ramp counter reset and field sync is 0.92 ms. Now, 0.92 ms − 416 μs = 504 μs, so if a 504 μs delay were interposed between the Ch.1 switch phase monostable (ramp reset pulse source) and the input of the Ch.1 flip-flop, then the latter would toggle 504 μs after the ramp counter had reset, leaving a further 416 μs before the field sync pulse. The figure of 504 μs is a practical one obtainable from the logic output of the ramp counter.

It can be seen from Figure 4.29 (bottom) that the logic output is taken from the ramp count of 0100011000 which equates to a decimal count of 280: 280 × 1.8 μs 504 μs. From counter reset, then, 504 μs will elapse before 0100011000 is reached to trigger the drum flip-flop – the remaining 416 μs is equal to 6.5 lines so that the flip-flop edge occurs precisely 6.5 lines before the field sync pulse. Thus is eliminated another capacitor timing circuit and attendant pre-set pot. This arrangement does extend the Ch.2 period of the drum flip-flop by a millisecond compared with Ch.1 period; mark/space becomes 20.5 ms/ 19.5 ms in both record and replay, but the error on replay is compensated for by correct setting of the play back switch phase adjustments.

Further uses of frequency discriminators

In an earlier section of this chapter (Dual-loop servos) we saw that modern servos embody two loops, one for phase control and another for speed regulation. The heart of the speed control loop is a frequency discriminator; early versions were outlined in Figures 4.15 and 4.21. Later designs incorporate more circuit functions within purpose-designed ICs, using a minimum of discrete peripheral components; servos based on these LSI ICs are characterised by sophisticated features and a low total component count. Typical of such a servo is the discriminator configuration shown with its timing waveforms in Figure 4.32.

An input pulse initiates the ramp in a sawtooth generator, only to be reset by the arrival of the next input pulse, and so on, so that the sawtooth period is the same as the input pulse period. The slope (rate of rise) of the ramp is governed by external RC components, and with the slope thus fixed the final voltage level achieved by the ramp (at the moment of reset) is dependent on the ramp-run *time*, i.e. the period of the input pulse. The sample and hold section samples the ramp voltage at its maximum level – the moment of reset – and stores it in capacitor Cs. Plainly, then, the stored charge in Cs is proportional to input pulse period, and a fall in input frequency (long

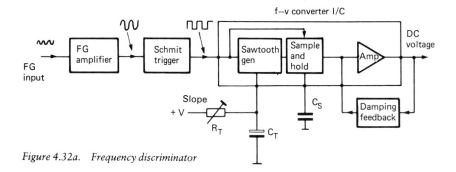

Figure 4.32a. Frequency discriminator

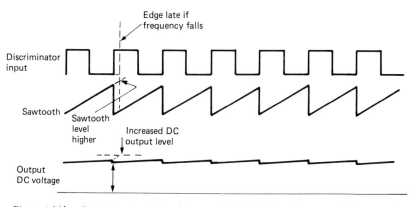

Figure 4.32b. Discriminator timing diagram. The broken line illustrates the increased DC output if the input frequency falls

period) will give rise to a high voltage at Cs and vice-versa, see Figure 4.32b. The circuit thus functions as a frequency to voltage (f-v) converter in which output voltage is inversely proportional to input frequency. If the input is taken from a motor's FG section, and the motor should slow down, the resulting increase in output voltage can be fed via an amplifier to the motor to compensate. The effect is to maintain constant motor speed, and that speed can be set by adjustment of the external RC time constants. While variable resistors (R_T in Figure 4.32a) are commonly used to set the 'free run' speed, pre-set voltages can also be introduced by switching transistors.

For drum and capstan servos the f-v convertor is used as the frequency discriminator for speed control. It may also be utilised in the reel motor servo loop to maintain correct visual search speed; here the input would be off-tape control track pulses. Where auto drum speed correction is provided for search mode the f-v convertor can be used to control the drum motor for maintenance of correct (15.625 KHz) off-tape line frequency. We have already seen that a 10× forward speed (cue) mode requires a drum speed increase of about 5%, and 10× speed reverse search (review)

mode a drum speed reduction of about 6%. Use of an f-v convertor enables this correction to be done automatically, eliminating the need to provide and set up trimming potentiometers.

Drum servo application

Figure 4.33 shows a more advanced drum servo system. In record mode the reference input is of course field sync, suitably divided by two, and the feedback signal consists of drum PG pulses – conventional practice for a drum servo. The speed control loop is based on an f-v convertor, taking a signal at 1.6 KHz from the drum FG. The only preset control (drum discriminator) enables the pulse position to be correctly set. Either an analogue or a digital servo system may be employed here, and externally there need be little difference between them; let us examine the operation of the discriminator control in terms of an analogue system.

Although the drum discriminator preset is within the *FG* loop its effect is to set the pulse position on the ramp. Imagine the sample pulse sits too low on the ramp. Adjustment of the discriminator control increases the f-v convertor

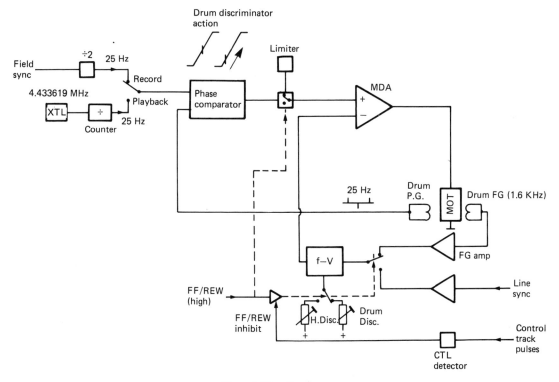

Figure 4.33. Drum servo

output voltage to the *inverting* input of the motor drive amplifier. MDA output voltage falls as a result and the motor slows down, retarding the arrival of PG pulses at the phase comparator (sample and hold) block. Their later arrival places them further up the ramp and increases the drive level to the MDA's non-inverting input. Pulse position on the ramp will continue to rise until the 'slowing' effect of the FG loop is balanced by the increased drive from the PG loop, arising out of the higher position of the pulse on the ramp. Hence the PG loop compensates for small variations in the FG loop, enabling the latter's drum discriminator preset to correctly position the pulse halfway up the ramp. An analogy can be seen in the phase locked loop of a TV flywheel line sync circuit, where small adjustments of the horizontal hold control will shift the picture laterally; if too large an offset is attempted control is lost at the point where the loop unlocks and all synchronisation disappears.

Reverting to Figure 4.33, the playback reference is a 4.433619 MHz crystal, counted down to 25 Hz. The drum locks to this to maintain constant speed and phase.

When visual (picture) search is selected, either forward or reverse, several switching functions take place. The phase control output is inhibited and replaced by a fixed voltage from the limiter (see Figure 4.20 for limiter); the input to the f-v convertor is switched to off-tape line syncs; and the drum discriminator preset is replaced by a 'horizontal discriminator' control. Since we are anticipating a change of FG tone from 1.6 KHz to 15.625 KHz (factor of ten) it is reasonable to expect the horizontal discriminator preset to present a resistance of about one-tenth that of the drum discriminator pot. In search modes, only the FG loop is operational, working to hold the drum speed at a point where off-tape line syncs have the correct frequency of 15.625 KHz. A safety function is also incorporated, monitoring control track pulses: if they disappear it will be because a blank section of tape is passing. In these circumstances the input to the f-v convertor reverts to the FG signal as a temporary reference until pictures reappear along with the control track.

Capstan servo application

The principles described in the last section are also applicable to a capstan servo, and the similarity between Figures 4.33 and 4.34 is immediately

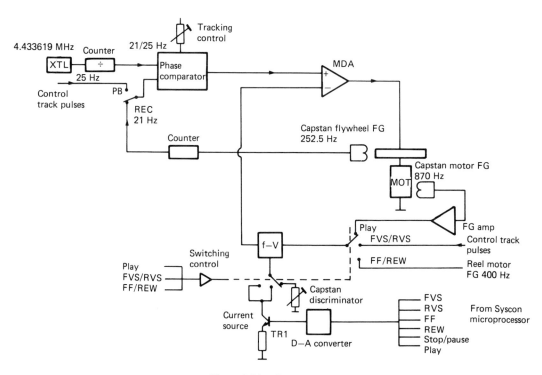

Figure 4.34. Capstan servo

obvious. Starting at the top left-hand corner of Figure 4.34 (capstan servo) the reference in both record and playback is the counted-down output of a 4.43 MHz crystal. During record, feedback from the rotating device comes in the form of a 252.5 Hz tone from a flywheel FG, divided to 21 Hz in a counter, and matching the 21 Hz output from the 'programmable' reference divider. In this machine the capstan servo runs at 21 Hz in record and 25 Hz in playback for compatibility with NTSC machines – Table 4.1 gives details. In playback this capstan servo is required to stabilise the tape speed *and* phase lock it to the recorded signal, and the feedback path from the rotating device is a little more obscure! In fact it comes via the tape ribbon itself in the form of replayed control track pulses. The control track pulses are written onto the tape during record at the rate of one per two video tracks, and their physical positioning on the tape has a fixed and known relationship to the positions of the video tracks themselves. In playback their timing is also indicative of the rotational phase of the capstan shaft, so can be used as a source of feedback from the rotational device. Tracking-correction phase shift is applied to the capstan servo, then, and with the drum servo working autonomously (i.e. fixed speed and phase), any phase shift with respect to it is corrected to achieve proper tracking in the face of a range of prerecorded tape tolerances.

The capstan servo f-v converter plays a major role, primarily for speed control as described earlier. An 870 Hz FG signal from within the motor provides the input to the frequency discriminator. If visual search is selected during playback the f-v input is switched to control track replay, and will operate to maintain an input frequency of approx 225 Hz, corresponding to nine times normal control pulse rate of 25 Hz. As we have already seen, this calls for a steeper slope in the ramp generator within the f-v convertor, and this is 'programmed' by the components towards the bottom of Figure 4.34. A *D-to-A* (Digital to Analogue) converter examines data from the system control microprocessor to produce a specific output voltage for each transport mode. This is applied to the base of constant-current generator TR1 whose collector current sets the ramp slope angle to the appropriate degree. Taking the visual search example given above, imagine that 'cue' is selected. A high binary number will be presented to the D-A converter and its correspondingly-high output voltage is presented to TR1 base, giving rise to a high collector

current. This rapidly charges the ramp-forming capacitor to generate the required steep slope.

In fast forward or rewind modes the same circuit alters the ramp-capacitor's charge rate to a new value which allows an FG signal from the take-up or supply turntables to be maintained at 400 Hz. Each turntable produces 48 pulses per revolution, so 400 Hz corresponds to 8.33 revs/sec. In fast forward mode pulses are taken from the feed spool, whereas in rewind they come from the takeup spool. By thus controlling the speed of the spool from which tape is being taken the fast transport speeds are limited to a steady value and the winding tension held steady, avoiding tape bunching or *cinching* on the receiving spool. A further advantage is that during sudden braking to terminate search mode (i.e. counter memory operation) the tape is subjected to minimum physical stress.

Summary

The use of frequency discriminators in advanced servos enables the designer to have safer control over tape handling, and allows for more control over the various conditions which are to be found in multiple function video recorders.

LSI servo IC

Rotational information from the videohead drum in Figure 4.35 is sent to the drum pickup input (PG) pin 22 as pulses or sometimes a 25 Hz square wave signal as the drum PG pulses are squared in a schmit trigger circuit externally to the IC. There is also the drum FG signal for speed control entering the IC on pin 21 at a frequency of 1500 Hz. This is integrated in the drum speed detector to produce the drum speed error signal for the pulse width modulator on pin 41.

Pin 19 is a head select input for 2 or 4 head use and is set to a fixed mode. On pin 17 is the output of the drum flip flop signal whilst the phase shifted FM audio head switching flip flop is on pin 18.

In record composite syncs are applied to pin 10, horizontal syncs are removed and vertical syncs are divided by 2 to 25 Hz and are used to reset the reference counter. In the absence of V sync, such as playback, the reference counter outputs 25 Hz pulses counted down from the reference input on pins 4 and 3. These are shown as a crystal oscillator, but are often 4.43 MHz from the colour circuits.

The output of the reference counter is used as the drum servo phase comparator reference whilst the drum flip flop provides the second, variable, input.

Drum PG pulses from pin 22 set a monostable with a variable delay connected to pin 20 which is used to set the 6H switch point position in playback. A set/reset bistable is used to obtain the drum flip flop on pin 17, there is an inbuilt 2H delay for dual speed versions to compensate for the 2 line shift between LP and SP video heads. Internally the flip flop signal is used on the drum phase detector and the drum phase error is on pin 42 as a 1.1 MHz pulse width modulated signal.

Both the drum speed and phase pulse width modulated signals are integrated to DC error signals and combined as the drum error signal in an adding amplifier before being applied to the drum motor. Their combined range of operation is shown in Figure 4.24.

Vertical pulse generator

This part of the circuit within the IC produces the synthesised vertical sync pulse which is used in still, slow and visual search modes. An inverter is used between LP and SP modes to compensate for the 180° phase shift between LP and SP video heads which are mounted opposite each other (see Figure 7.26).

Two delay monstables form the vertical syncs with one monostable being adjustable for vertical jitter, as the vertical lock control on pin 16. A vertical width counter determines the actual width of the synthesised vertical sync pulse on pin 13.

Record CTL track

An output from the reference counter is used to construct the control track recording signal. After delay by the CTL delay monostable the record duty cycle counter determines the 28/12 ms mark space ratio for recording out of pin 30.

Capstan servo

The record CTL signal is also used as the reference for the capstan phase comparator with the variable input coming from the capstan FG signal on pin divided by 20. As the FG signal is 500 Hz

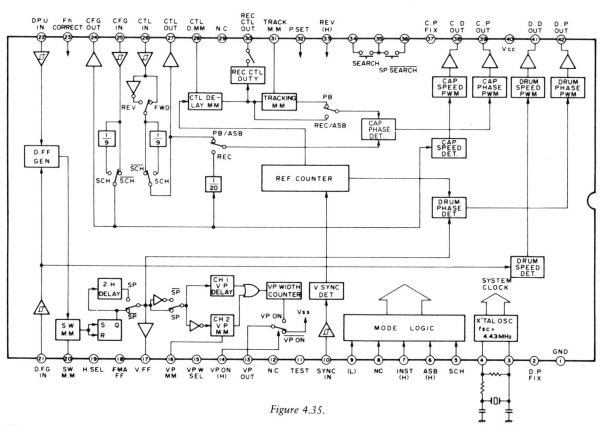

Figure 4.35.

division by 20 results in 25 Hz. Capstan FG signal is also integrated in the capstan speed detector for speed control similarly to the drum.

In replay the drum phase control consists of the drum PG pulses and a 25 Hz crystal reference, counted down in the reference counter from the 4.43 MHz oscillator.

The capstan phase detector input is replayed control track from pin 26 via the switching and tracking monostable on pin 31. Whilst the reference for the phase detector is also from the reference counter output at 25 Hz counted down from the 4.43 MHz input on pins 3 and 4.

The special facilities accommodated within the IC are phase locked picture search and assembly editing.

For picture search operation at 9x play speed the capstan FG input is divided by 9 to run the capstan speed at 9x play. Loose phase lock is achieved by comparing the replayed control track, also divided by 9, to the reference counter with an extra division by 9 inbuilt into the programmable counter, this stabilises the noise bars on the TV screen.

Assembly editing is achieved by the ASB switch positions where it is required to lock the capstan servo to the incoming signal for a short while (approximately 500 ms; see assembly editing). It enables the capstan phase detector to reference from incoming syncs (REC/ASB) and replayed control track (PB/ASB) for the short run in period before switching over to full record mode.

Pins 34 to 36 control the capstan servo gain for visual search while mode control logic from the main micro computer controls the functions on pins 5–9. Pins 2 and 37 fix the drum and capstan discriminators for test purposes, pin 33 controls the capstan motor reverse operation.

5

COLOUR SYSTEMS

In Chapter 1, the bandwidths of the various systems are illustrated. One of the most prominent frequencies is that of the colour subcarrier record/replay 'colour under' carrier.

The term 'colour under' refers to the shift of the colour carrier from the top end of the band at 4.43 MHz to a frequency below that of the luminance FM band. The frequencies are around 600 –700 kHz and are different for all systems, i.e. Philips N1500 is 562.5 kHz, U-Matic 685 kHz and VHS 627 kHz. The colour is phase and amplitude modulated onto the colour under carrier, as it was on the original subcarrier. A technique is used to add and subtract an intermediate carrier in order to convert from a modulated subcarrier to a modulated colour under carrier for recording.

In replay the same intermediate carriers are utilised to convert back from colour under carrier to subcarrier and provide compensation for phase and jitter errors caused by the mechanics and physical movement.

Balanced modulator

A balanced modulator is a three terminal device (Figure 5.1). Two inputs are frequencies F1 and F2. Appearing at the single output are both the

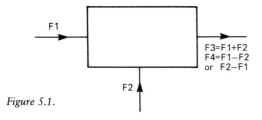

$$F3 = F1 + F2$$
$$F4 = F1 - F2$$
$$\text{or } F2 - F1$$

Figure 5.1.

sum and the difference of the inputs. That is to say one output F3 will be F1 + F2 and the other output will be F4, either F1 − F2 or F2 − F1, whichever is the greater.

In a working system, for example the Philips N1500 is a basic format, F1 is the 4.43 MHz colour signal, F2 is the intermediate frequency of 4.99 MHz, F3 is 9.42 MHz and F4 is 526.5 MHz.

F4 is the wanted signal, being the recorded colour under signal, and it is separated from F3 via a low pass filter.

The balanced modulator is the basis of all colour under systems, and it is utilised to add and subtract intermediate carriers. The output from the balanced modulator is selected by high, or low pass, or bandpass filters. The most important point in the colour under system is that the colour under carrier must be phase locked and referenced to the luminance signal. If the colour under carrier is not locked in phase to the line syncs then replay correction would not be possible; this would result in changes in hue.

There are various delay lines used in the circuitry. Due to the additional processing that the colour signal has to undergo it suffers from signal delay, consequently the luminance channel has additional delays in the form of a delay line to compensate. In order to achieve phase lock of the colour under carrier, intermediate carriers are used which are derived from line syncs in an AFC loop.

N1500 colour under system

The Philips N1500 colour under system is outlined in Figure 5.2. A 'base carrier' of 562.5 kHz is derived from a phase lock loop. Within the loop a voltage controlled oscillator runs at 562.5 kHz; the output is divided by 36 to f'_h (15.625 kHz), f'_h is compared to incoming line syncs from the video signal, f_h, in a phase comparator. The output error signal from the phase comparator controls the voltage controlled oscillator and thus keeps f'_h and f_h in phase. The 562.5 kHz output from the oscillator is therefore also in phase with line syncs.

The phase locked base carrier is used as one input to balanced modulator 1. A second input comes from a 4.43 MHz oscillator. This need not be phased locked by anything, but on the N1500 it is kept on frequency by gating the colour burst, similar to a standard PAL decoder. An output from balanced modulator 1 is obtained via a 4.99 MHz filter; this is the intermediate carrier for balanced modulator 2, and it also contains the phase locked characteristic. In modulator 2 the incoming colour signal, amplitude and phase modulated with colour information, is mixed with the intermediate carrier, which is locked in phase, a pure carrier.

The outputs from balanced modulator 2 will be 9.42 MHz and 562.5 kHz, both amplitude and phase modulated with colour information. A low pass filter selects 526.5 kHz. Note that, as the output from a balanced modulator is both sum and difference of the inputs, it follows that if one input is a modulated carrier and the other is not, then the resultant output will also be modulated.

In the replay situation the frequency of the replayed colour undercarrier will vary due to very small acceleration and deceleration of the scanning video head, tape path movement and tape tension variation. It is necessary to remove these timing errors, often referred to as jitter. The colour under signal is converted back to 4.43 MHz in replay by reversing the actions of balanced modulator 2. The intermediate carrier has to be the same for record and replay, but there are slight differences in its derivation. First, and least important, is that the 4.43 MHz carrier into balanced modulator 1 is a free running crystal oscillator. Second, and most important, is that line syncs into the phase locked loop come from replayed video.

Phase and jitter reduction

Assume that the replayed colour undercarrier has a jitter, J. As the line syncs into the phase locked loop are from replayed video they will also have a jitter component, J. Following through the system, the base carrier 562.5 kHz will also contain J, and also the intermediate carrier, $(4.99 \text{ MHz} + J)$. In balanced modulator $(562.5 \text{ kHz} + J)$ is subtracted from $(4.99 \text{ MHz} + J)$, resulting in 4.43 MHz output and the cancellation of J. J can be a change in frequency in the region of a number of cycles shift. The acceleration and deceleration of the video head drum or changes in tape speed can be such that the replayed colour undercarrier could be, for example 566 kHz. In this case line syncs would increase by the same proportion and would result in an increase in frequency of the phase locked loop output to 566 kHz. The colour subcarrier output would still be the correct 4.433619 MHz. J can also be a change in phase, that is a shift of less than one cycle, a phase correction would then be applied by the same technique.

Unless a fault occurred, a videorecorder cannot replay a programme with a different hue or colour due to the corrections applied. However, in the N1500 system the balanced modulator 1 input 4.433 MHz could be off frequency and so adjustment is provided for this.

Setting up procedure is given in the service manual using a DC reference voltage. Fine adjustment in replay is best done by comparing direct or E/E flesh tones with replayed flesh tones and adjusting as necessary.

Colour crosstalk

Referring to the magnetic track patterns given in Chapter 1 it can be seen that the video tracks are spaced apart by the guard band sufficiently to prevent any crosstalk between tracks in the N1500 system. When the N1700 system was produced with the magnetic tracks next to each other luminance crosstalk in replay was eliminated by azimuth offset between the two video head gaps. Whilst this azimuth slant worked well at high frequencies it had little effect at low frequencies, the low frequency in question being the colour undercarrier at 562.5 kHz.

The N1700 track pattern is offset 1½ lines between adjacent field tracks and as such, the phase of the colour burst is the same between any two lines that lie next to each other in adjacent

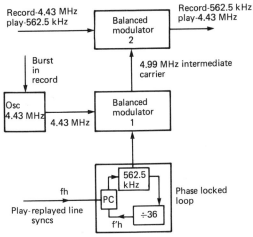

Modulator 1
4.433619 mHz $\mp$ 562.5 kHz = 4.996119 MHz

Modulator 2
Record   4.433619 MHz $\pm$ 4.996119 MHz = 562.5 kHz
Play   562.5 kHz (+J) − 4.996119 MHz (+J) = 4.433619 MHz

Figure 5.2.   N1500 colour under system

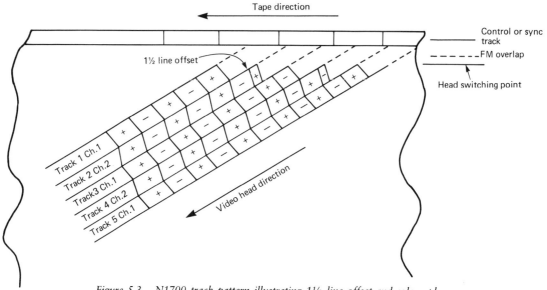

Figure 5.3. N1700 track pattern illustrating 1½ line offset and colour phase

field tracks. This means that if head Ch.1 is replaying its own magnetic track and wanders off track any colour crosstalk picked up from an adjacent video track belonging to head Ch.2 will be in phase. The colour crosstalk is then additive rather than subtractive. Luminance crosstalk is not reproduced due to the azimuth slant. The theory that what the eye does not see then the heart will not grieve over is applied. Any given single line will not have much change in its colour content between a preceding field and the present one being replayed, or the one to follow. Crosstalk interference in the N1700 is fairly small and can be coped with in this way.

There are no correction techniques required for colour crosstalk compensation in the N1700. The record/replay processing is the same as for the N1500 except that ICs are used in place of discrete components.

In the VHS, Beta and V2000 formats, the 'packing density' of video information onto the magnetic tape is much higher than in previous developments. It is necessary in these more common systems to provide for a higher degree of colour crosstalk compensation, or rather the aim of the system is colour crosstalk cancellation. The point to bear in mind, before reading about the individual approaches, is that although each system appears different, in fact they are very similar. Each system utilises a 'comb filter' to eliminate the crosstalk produced by the replay heads. The comb filter is designed around a two line delay line. The

basis for a two line delay is the PAL signal itself. The PAL R − Y signal is phase reversed each line and hence is always 'in phase' over every two lines.

The idea behind all of the systems is to lay a magnetic track in the record mode such that any crosstalk reproduced in replay appears in antiphase over every two lines. Consequently if the third line is added to a delayed first line the wanted PAL signal will be in phase and the unwanted crosstalk will be in antiphase.

Operation of the comb filter

A comb filter consists basically of a tuned delay line followed by an addition network. The frequency response of a comb filter, as seen on a spectrum analyser is as in Figure 5.4. The centre frequency f_o has sidebands of multiples of f_o, that is $2f_o$, $4f_o$ etc. If the comb filter is placed in series with a bandpass filter then the upper and lower frequency bands are eliminated by the tuned bandpass filter, assuming of course that it is also tuned to f_o. The total frequency response is the single

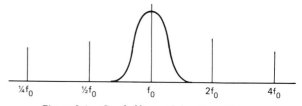

Figure 5.4. Comb filter with bandpass filter

90

output f_o from both filters in series, this gives a very sharp filter and when the centre frequency f_o is 4.433619 MHz it can be used to eliminate sidebands of crosstalk which are deliberately produced.

In the video recorder, where f_o is 4.433619 MHz, the delay line is a two line period delay, the signal is fed through a 4.43 MHz bandpass filter and the direct and delayed outputs are combined to achieve the required response.

The principle for replayed colour crosstalk cancellation is to develop a magnetic track pattern which will produce crosstalk between tracks in replay which is in antiphase every two lines. In the VHS system, when Ch.2 head records the colour undercarrier the phase of this carrier is rotated by $-90°$/line, or $-180°$ for every two lines with respect to line syncs.

The Betamax system records Ch.1 with the colour carrier phase retarded by $-45°$/line and Ch.2 head advanced by $+45°$/line, with respect to line syncs. The result is a phase rotation of $-90°$/line of Ch.1 with respect to Ch.2, or as in VHS, $-180°$ phase reversal every two lines.

Taking the VHS system one step further, it is not too difficult to say that if we are retarding the phase of Ch.2 head by $-90°$/line or $-180°$/two lines, then we are subtracting one whole cycle every four lines. At a line rate of 15.625 kHz/ second we are subtracting 1 Hz every four lines, that is 15 625 divided by 4 resulting in 3 906 whole cycles subtracted each second, that is 3.906 kHz.

It can thus be argued that, for VHS with a colour undercarrier frequency of precisely

Ch.1 head and Ch.2 head respectively. If these are multiplied out:

$$(44 - ⅛) \times 15\,625\,\text{Hz} = 685.54688\,\text{kHz}$$
$$(44 + ⅛) \times 15\,625\,\text{Hz} = 689.45312\,\text{kHz}$$

The difference being 689.45312 kHz − 685.54688 kHz = 3.90624 kHz.

Thus we find that Betamax has the same carrier frequency difference as VHS, 3.906 kHz. This difference frequency of 3.906 kHz is directly related to a vector rotation of $-90°$/line of one colour undercarrier, Ch.2 video head, with respect to the other, Ch.1 video head.

A vector rotation of $+45°$/line for Ch.1 colour undercarrier with respect to line syncs and a vector rotation of $-45°$/line for Ch.2 colour undercarrier will be equivalent to 90°/line of one with respect to the other.

The delay line

Let us consider the case of a 4.433619 MHz signal replayed into the comb filter delay line. The delay is 127.88645 μs, but how is this figure arrived at?

The standard delay line in a colour television PAL decoder is not exactly one line delay, 64 μs, it is slightly less. The colour subcarrier is 4.433619 MHz so the period of one cycle (Hz) is 0.2255493 μs duration and if you divide the delay line time (64 μs) by this figure the answer is 283.75.

This is the number of whole cycles of 4.43 MHz signal that would be present inside the delay line at any one time (Figure 5.5). In order for the delay

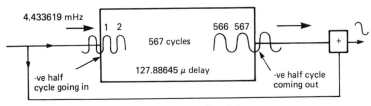

Figure 5.5. Two-line delay

626.952 kHz for Ch.1 head, the resultant colour undercarrier for Ch.2 head will be 3.906 kHz lower, that is 623.045 kHz.

Back to the Betamax system, which has a phase shift of $-45°$/line for Ch.1 head and $+45°$/line for Ch.2 head. There are two formula for the derivation of colour undercarriers. They are: $(44 - ⅛)f_h$ and $(44 + ⅛)f_h$, where f_h is line frequency, for

line to have the necessary 180° phase reversal between input and output there would have to be an odd half cycle, not three quarters (0.75) so the spare quarter cycle has to be done away with. Taking then that the content is to be 283½ cycles, this shortens the delay to 63.943226 μs.

A two line delay with no phase reversal between input and output would have a whole number of cycles content at any one time, that is 567 cycles. In order to have a content of 567 cycles the delay for a two line delay line will have to be 127.88645 μs. Hence as in Figure 5.5 the output is in phase with the input, when summed in the adder network the output is a 4.43 MHz signal.

Consider now that if Ch.1 head were replaying and it was picking up Ch.2 head's colour signal then this unwanted pick up is crosstalk. The reconversion system would convert Ch.1 head colour under carrier to 4.433619 MHz. Ch.2 head's crosstalk would also be converted by the same process so it would produce a carrier which was down by 3.906 kHz, that is, 4.429712 MHz. In the same manner, Ch.2 head replaying would convert the colour under carrier to 4.433619 MHz. Ch.1 head crosstalk would also be reconverted and in this instance it would be higher by 3.906 kHz, that is, 4.437526 MHz.

This information means that the comb filter can have either one of two crosstalk frequencies applied to it. These crosstalk frequencies are the same for VHS and Betamax. By calculating the number of cycles of carrier present in the delay line at the crosstalk carrier frequencies the results are as follows: 566½ for 4.429712 MHz and 567½ for 4.437526 MHz. The odd half cycle in both cases indicates that a 180° phase reversal occurs between the input and output of the delay line. Consequently, as shown in Figure 5.6, when a positive half cycle is entering the delay line, a negative half cycle is coming out. Mixing the input

Colour crosstalk frequencies, replayed by either video head, will be 3.906 kHz either side of the centre frequency. These frequencies will undergo phase inversion and will therefore be cancelled in the additive mixer at the delay line output.

A frequency approach to crosstalk cancellation is the more usual description given by Betamax systems. The VHS system is more usually explained in vectorial terms, but the two can be related as they are similar.

In Figure 5.7 two adjacent magnetic tracks are shown, the phase of the colour burst for Ch.1 video head is shown recorded as a colour under carrier, its phase unaltered by the recording process. Next to Ch.1 video head track is Ch.2. The phase of the colour signal in this track is rotated in an AFC loop by −90°/line, the rotation is clockwise.

Line 'n' is taken at random when the vector rotation for Ch.2 video head is at 0°.

In line (n + 1) the vector for Ch.2 is the same as for Ch. 1 except that it is rotated clockwise by 90°.

In line (n + 2) the vector for Ch. 2 is the same as for Ch.1 except that it is rotated clockwise by 180°, or inverted.

In line (n + 3) the vector for Ch.2 is the same as for Ch.1 except that it is rotated clockwise by 270°.

The pattern is repetitive throughout the recorded track. It sets up a recognisable vector pattern for Ch.2 video head which can be called

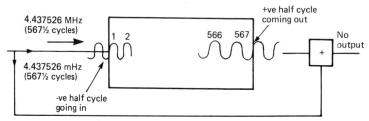

Figure 5.6.   Two-line delay with 4.437526 MHz crosstalk

and the delayed output together in the additive network results in the cancellation of the crosstalk carriers at the output of the comb filter and hence the elimination of crosstalk.

The vectorial approach to VHS

We have already seen that in the comb filter, consisting of a two line delay period, a subcarrier of 4.433619 MHz will pass through without being inverted in phase, as this is the centre frequency.

'inverted line pairs'. Consider the two line period delay line, where at any given instant the signal out of the delay line is that which went in two lines earlier. The additive mixer at the output of the delay line will combine two signals that are two lines apart. By referring to the replay track patterns it can be seen that the burst phase for Ch.1 video head replay will always be in phase over any two lines. Ch.2 signal picked up by Ch.1 head as crosstalk will always be in antiphase over any two line period.

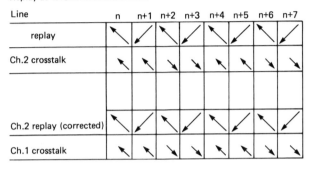

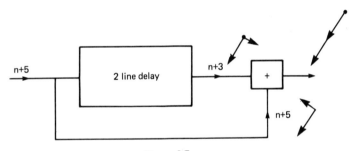

Figure 5.7.

It is essential for *any* colour record/replay system to arrange for the replayed colour crosstalk to be replayed in the 'inverted line pair' pattern in order to cancel in a two line period delay line comb filter. If we take for example the replay of line n + 5, it is fed directly to the additive mixer, where it is mixed with a line appearing out of the delay line. This line is one which was replayed two lines earlier, that is, n + 3. The wanted signal is in phase and is vectorially added; crosstalk signals over two lines are in antiphase and will therefore cancel.

It is easy to see that when Ch.2 colour under signal is recorded whilst rotating it by −90°/line it sets up the inverted line pairing pattern. In replay we can say that the Ch.2 wanted signal is corrected by adding back the phase rotation, +90°/line. In doing this the Ch.1 video head signal is replayed as crosstalk, and as it was not phase modified during record it is affected by the correction process.

The correction process of 'adding 90°/line' to Ch.2 replay will cause the unmodified Ch.1 crosstalk to be phase rotated by +90°/line. Crosstalk will be fed to the delay line in the inverted line pairing pattern. It can be proved that by rotating the vectors by +90°/line that the swinging burst will be converted to inverted line pairs, and cancelled. It can be seen later that the 90°/line is added back in modulator 2 by a mathematical process and that the Ch.1 crosstalk in a Ch.2 replay situation is modified by −90°/line.

93

The VHS system

The VHS colour recording system is shown in block diagram form in Figure 5.8. The easiest approach to the colour system is to start at the bottom of the system 'tree'.

The 'base frequency' is generated in a phase locked loop as previously shown in the N1500 562.5 kHz example. This frequency is 626 kHz – note that this is not the same as the colour under record frequency of 626.9 kHz, the shift being produced in balanced modulator 1, by 4.435571 MHz instead of 4.433619 MHz.

AFC loop

The main oscillator is a 2.5 MHz voltage controlled oscillator and the output frequency is divided by four in a counter. Tapped off the counter is 625 kHz in its four phases, 0°, 90°, 180° and 270°; further division of 625 kHz by 40 in an 'additional' or 'second' counter chain results in signal f'_h which is 15.625 kHz. f'_h is compared with incoming line syncs, f_h, from the video signal, in a phase comparator. An error voltage from the phase comparator is then fed back to the voltage oscillator to maintain f'_h in phase with f_h.

The phase locked loop is used to produce the base frequency of 625 kHz, phase locked to line syncs. An additional input to this loop is a 25 Hz flip-flop signal coming from the drum servo containing the information as to which video head is scanning. It is 'high' for Ch.1 and 'low' for Ch.2, and determines the phase selector operation.

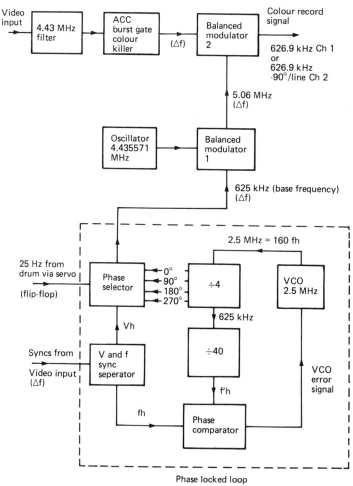

Balanced modulator 1 625 kHz + 4.435571 MHz = 5.060571 MHz
Balanced modulator 2 5.060571 MHz - 4.433619 MHz = 626.952 kHz

Figure 5.8. Block diagram of VHS colour record system

In the 'high' mode the phase selector is held at 0 for Ch.1 video head, in the 'low' mode the phase selector 'rotates' and produces 625 kHz in each of the four phases sequentially. It is an interesting point to note that if a frequency counter is connected to measure 625 kHz rotating by −90 line at this point it will read 2.5 MHz. This is because the 625 kHz is in four switched phases; it is not a continuous waveform, but a four segment switched signal, hence the reading of 625 kHz × 4 = 2.5 MHz.

The 625 kHz is then fed to balanced modulator 1 where it is added to 4.435571 MHz, produced from a crystal locked voltage controlled oscillator. The resultant sum output of 5.06 MHz is filtered by a bandpass tuned network. At this point Ch.2 frequency becomes a continuous sinewave carrier and is 3.906 kHz below Ch.1 frequency. Ch.1 is 5.060571 MHz and Ch.2 is 5.056664 MHz; these can now be measured, but they are alternating at a rate of 25 Hz, so before measurement the phase lock loop has to be held in Ch.1 mode or Ch.2 mode by fixing the flip-flop input at 'high' or 'low'. The intermediate frequency of 5.06 MHz is fed to balanced modulator 2, and the other input to this balanced modulator is the colour modulated 4.43 MHz colour input.

Input video is filtered and level controlled in an ACC circuit (automatic colour control) and fed to a record colour killer to balanced modulator 2. This modulator is used in the difference mode and so the colour input is subtracted from the intermediate frequency of 5.06 MHz, the output of the balanced modulator being 626.9 kHz. It is selected by a low pass filter and fed via buffer amplifiers to the recording amplifiers.

The colour under carrier for VHS Channel 1 video head is 626.952 kHz. Channel 2 video head colour under carrier is 626.9 kHz − 90°/line; this is equivalent to subtracting 3.906 kHz as previously proved, hence the frequency is 623.046 kHz for Ch.2 colour under carrier.

An additional characteristic of the colour record system is that of differential phase correction. The videorecorder will be recording signals received from transmissions that are subject to phase distortion. It is well known that within a television receiver the PAL decoder will compensate for transmission phase errors. If the colour recording circuits of a videorecorder did not also compensate then recorded hue variations would result.

The compensation is identical to the AFC action in the replay mode. If the colour signal input to modulator 2 contains a phase error Δf then it is reasonable to expect that the incoming line sync pulses will also contain Δf shift. This will result in the base carrier 625 kHz also containing Δf. Through balanced modulator 1 the phase error will carry onto the intermediate carrier 5.06 MHz. However, in balanced modulator 2 the incoming colour is subtracted from the intermediate carrier and the phase error Δf will cancel.

The VHS replay system

AFC loop

Figure 5.9 illustrates the VHS replay system in block diagram form. In the replay mode the action of the phase lock loop is to produce the base frequency the same as in record. The only difference being that the syncs, f_h, to be compared in the loop, come from replayed video. The loop then locks to the replayed signal. This forms the basis for frequency correction. In a similar manner to the N1500 system these syncs will contain any jitter components $\pm J$, to be used to cancel jitter contained within the replayed colour undercarrier.

Sync $\pm J$ will result in (625 kHz $\pm J$) within the phase locked loop. In modulator 1: (625 kHz $\pm J$) + 4.435571 MHz = 5.060571 MHz $\pm J$. In modulator 2: (5.060571 MHz $\pm J$) − (626.9 kHz $\pm J$) = 4.433619 MHz. It can be seen that in balanced modulator 2 the $\pm J$ component is cancelled.

−90°/line replay correction

Contrary to the indications given in various service manuals, when Ch.2 video head is replaying, the phase selector within the AFC loop does *not* change direcoton of rotation. It is very convenient to say that as 90°/line is subtracted during record then it is added back during replay. The statement is correct in so much as during Ch.2 replay there is a correction of +90°/line but it is not from the AFC loop.

Consider that −90°/line is equivalent to subtracting 3.906 kHz; then +90°/line is equivalent to adding 3.906 kHz. If it is agreed that the phase lock loop runs at −90°/line during record then the base carrier produced is equivalent to 621 kHz, in round figures, which is correct. However if it is further argued that the phase lock loop runs at +90°/line in replay then the base carrier produced is equivalent to 629 kHz, again in round figures. This will produce an 8 kHz shift upwards in the replay colour carrier, and confirmation of this can be obtained by the record/play carrier figures.

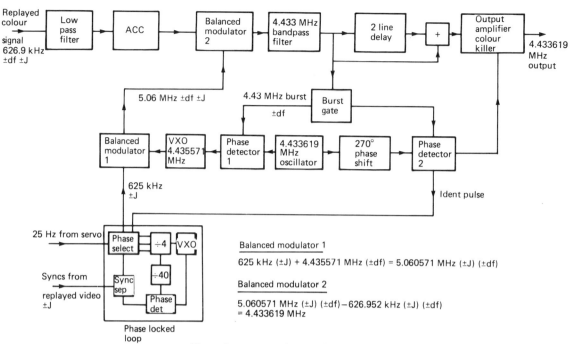

Figure 5.9.   VHS colour replay system

Therefore the phase lock loop, AFC circuit, cannot run at +90°/line in replay. The correction is carried out mathemtically by still running the phase lock loop at −90°/line. If you do not believe this then check the practical circuits and see if there are any rec/replay signals to the AFC loop which would instruct the phase rotator, in fact no such signal exists.

−90°/line replay correction

Looking at balanced modulator 1:

$(625\,\text{kHz} − [90°/\text{line}]) + 4.435571\,\text{MHz}$
$= (5.060571\,\text{MHz} − [90°/\text{line}])$

In balanced modulator 2 the replayed colour signal is subtracted from the intermediate carrier.

$(5.06057\,\text{MHz} − [90°/\text{line}]) − (626.9\,\text{kHz} − [90°/\text{line}])$

Expand the formula and this results in a sign change of the replayed signal:

$5.06057\,\text{MHz} − [90°/\text{line}] − 626.9\,\text{kHz} + [90°/\text{line}]$

The output from balanced modulator 2 is:

$5.06057\,\text{MHz} − 626.9\,\text{kHz} = 4.433619\,\text{MHz}.$

Crosstalk replay

Replayed crosstalk signal from adjacent Ch.1 video tracks will not contain any 90°/line switching.

$5.06057\,\text{MHz} − [90°/\text{line}] − 626.9\,\text{kHz}$
$= 4.433619 − [90°/\text{line}]$

The 90°/line component which is a 3.906 kHz shift will be eliminated in the comb filter.

Conclusion

There is a +90°/line signal in replay colour circuits to correct for −90°/line in record and it is derived in a subtraction process in balanced modulator 2.

Phase errors, that is errors of less than one Hz, are cancelled within modulator 2. This is only true if the rotation sequence of 0°, 90°, 180°, 270° is synchronised between record and replay, but it is *not*. It is therefore possible that a dropout in the tape could cause a replayed sync pulse to be missing and the phase lock loop will miss a stop. This refers only to a Ch.2 head replay.

In this instance the Ch.2 colour under signal replay *could* be replaying in the rotation sequence at 270°, whereas the phase lock loop may be in the 180° position; this then will give an error of 90° in the reconverted colour signal at 4.433619 MHz. Fear not though, as phase detectors 1 and 2 will come to the rescue.

Operation of the phase detectors

Colour burst, reconverted to 4.433 MHz, is gated from the replayed colour signal and fed to phase detectors 1 and 2. Also fed to these phase detectors is a 4.433 MHz reference carrier from a crystal oscillator, shifted by 90° to phase detector 2 (see Figure 5.10). The inputs to any phase detector should be shifted with respect to each other by 90° for the output correction voltage to be at 0° reference.

In the case of phase detector 1 neither input is shifted so the output correction voltage is centred about 90°. The correction range of phase detector 1 is ±90° either side of the 90° reference point, giving a total range of 180°. In phase detector 2 the reference carrier is shifted by 90° so that the output of this phase detector is 0 volts at 0°. This results in an output that is +ve for errors between 0° and 180° and −ve for errors between 0° and −180°. This output is a switching signal called the 'ident pulse' (360° to 180° = 0° to −180°).

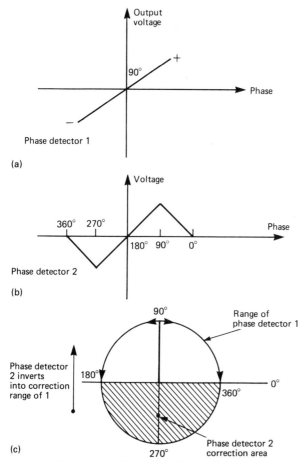

(a) Phase detector 1

(b) Phase detector 2

(c)

Figure 5.10. Operation of phase detectors

If a small phase error d*f* appears on the replayed colour signal and it is within 90° then the output voltage of phase detector 1, which is analogue, changes in level. This change is an error signal to a voltage controlled oscillator, crystal controlled at 4.435571 MHz. A crystal oscillator cannot change much in frequency but it can be 'bent' in phase. The phase error d*f* is then added to the intermediate frequency of 5.06 MHz.

In balanced modulator 2:

$$(5.060571 + \mathrm{d}f) - (626.9\,\mathrm{kHz} + \mathrm{d}f)$$
$$= 4.433619\,\mathrm{MHz}$$

In balanced modulator 2 the small phase error d*f* is cancelled.

We now have a phase cancellation loop from balanced modulator 2, through a 4.433 MHz filter, burst gate, phase detector 1, 4.435571 MHz voltage controlled oscillator, balanced modulator 1 and back to balanced modulator 2. Through this loop phase detector 1 can correct for any phase errors between 0° and 180°. Phase detector 2 will act if the error d*f* is more than 180°. If this happens the ident output of phase detector 2 will change state and this will travel to the phase lock loop. In the 625 kHz output path from the loop is a switching inverter and it can invert the 625 kHz every time the ident signal changes state.

The combined action of both phase detectors can result in 360° correction with respect to the 4.433 MHz reference oscillator, giving full correction. So if the phase rotator for a Ch.2 head replay gets out of step, phase detector 2 will invert the base carrier if it is more than 180° and then phase detector 1 will provide fine correction within ±90°, effectively bringing the replay phase rotator into synchronisation.

The phase correctors will correct for replayed errors as well as Ch.2 rotating phase errors; the resultant replayed and crosstalk cancelled signal is dependent on the 4.433619 MHz reference crystal for phase stability.

Full correction is applied to the replayed colour signal in three stages:

1. Frequency compensation by replayed line syncs into the phase locked loop.
2. Phase detector 2 to ensure that phase errors remain within the domain of phase detector 1.
3. Phase detector 1 maintaining correction within 180° referred to a crystal oscillator.

A further application of phase detector 2 is that of colour killer. If the ident pulse gets too repetitive then a time constant capacitor charges up and switches in a colour killer in the output amplifier. A fault in the colour record and replay circuits will result in one symptom – no colour! So start at the bottom of the block diagram and work your way up – you will find it.

Betamax colour system

As we have previously seen, the Betamax colour under system records two colour under carriers, one for head Ch.1 and another for head Ch.2. The difference in the two carriers is that they are rotating in phase with respect to each other by 90°/line, this being a frequency shift of 3.906 kHz.

The basic formula for Betamax is given as:

$$\cos(\omega ct + \phi) = \cos[(44 \pm \tfrac{1}{8})f_h t + \phi]$$

Simplified this gives the frequency for each head as:

Ch.1 head $= (44 - \tfrac{1}{8})f_h$
Ch.2 head $= (44 + \tfrac{1}{8})f_h$

Consider that 360° divided by 8 is 45, so Ch.1 head is rotating by $-45°$/line and Ch.2 head is

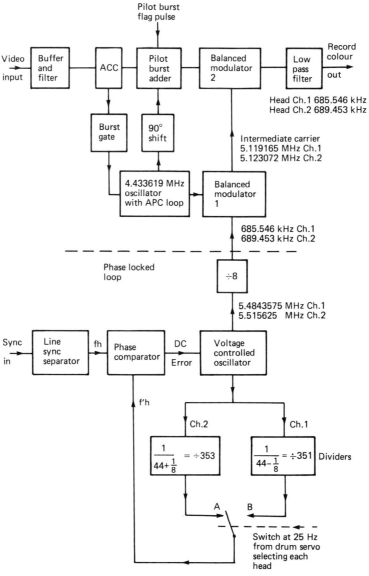

Figure 5.11. Betamax colour record block diagram

rotating by +45°/line. The −45° and +45° is with respect to line syncs, but the rotation of one carrier with respect to the other is 90°/line.

Given that f_h is line frequency, 15.625 kHz, the precise frequencies for each head can be worked out:

Ch.1 head = 685.546 kHz
Ch.2 head = 689.453 kHz

The difference frequency between the two is of course 3.906 kHz. Due to the relative shift between the two heads of 90°/line, any crosstalk picked up by either head will show the antiphase line pairing pattern shown for VHS and hence crosstalk will cancel.

Figure 5.11 shows the Betamax colour under recording system in block diagram form. The base frequencies for Ch.1 and Ch.2 heads are derived in the now familiar phase lock loop. A voltage controlled oscillator is running at 5 MHz and its output is divided by 8 in a counter. The counter output gives the base frequencies of 685 kHz and 689 kHz for each head. Note that these are the same as the colour record carriers, as the input to balanced modulator 1 is 4.433619 MHz.

Within the phase lock loop are two counters and these divide by (44 − 1/8) and (44 + 1/8), which is ÷351 and ÷353 respectively. These two outputs are selected sequentially by a switch driven at 25 Hz from the drum servo. The position of this switch indicates to the phase lock loop which video head is scanning the tape. The output from the selector switch will be f'_h and when it is compared to incoming line syncs, f_h, an error control signal is derived for the voltage controlled oscillator.

When the switch is in position A the division ratio is 353, so in order for f'_h to be in the same frequency and phase as f_h the voltage controlled oscillator is run at 5.515625 MHz.

When the switch is in position B the division ratio is 351; this time the voltage controlled oscillator is run at 5.4843575 MHz in order to phase lock f'_h and f_h. The phase lock loop will then be changing frequency at a rate of 25 Hz changing the output frequency of the voltage controlled oscillator at the same rate. Through the divide by eight counter the two base frequencies emerge also changing frequency between 685 kHz and 689 kHz for each head at a rate of 25 Hz.

In balanced modulator 1, 4.433 MHz is added to the sequenced base frequencies; it is the output of an APC (automatic phase control) oscillator and is consequently phase locked to the incoming burst

for extra stability. This is most useful if the input is from a camera or another source where the colour encoder 4 MHz oscillator is not locked to line syncs, i.e. it is a free running source of colour carrier. Another use of the APC loop is to provide for the system a phase locked carrier which can be gated onto the record signal for replay reference purposes.

In the Betamax system the swinging colour burst is not considered to be suitable for replay phase correction and so another burst of 4.433 MHz called the pilot burst is used. The pilot burst is gated onto the colour signal in a position which is the same as line syncs pulses (see Figure 5.12).

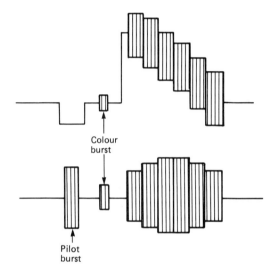

Colour burst

Pilot burst

Figure 5.12. Position of pilot burst

After replay colour correction when replayed colour is mixed back with replayed luminance the line sync pulse will wipe the pilot burst out. In the VHS system the gated burst to the phase correctors required 90° phase shift if the error signal had to be at 0° phase. In the Betamax system this is achieved by shifting the pilot burst by 90° prior to record processing. The pilot burst is then a gated 'chunk' of the APC loop output which is shifted by 90°.

Balanced modulator 1 produces two intermediate carriers at 5.11 MHz and 5.12 MHz alternating at the 25 Hz head rate, these are passed on to modulator 2 where they are mixed with the incoming colour signal and pilot burst. In balanced modulator 2 subtraction takes place resulting in two sequenced outputs, 685.546 kHz for Ch.1 head and 689.453 kHz for Ch.2 head.

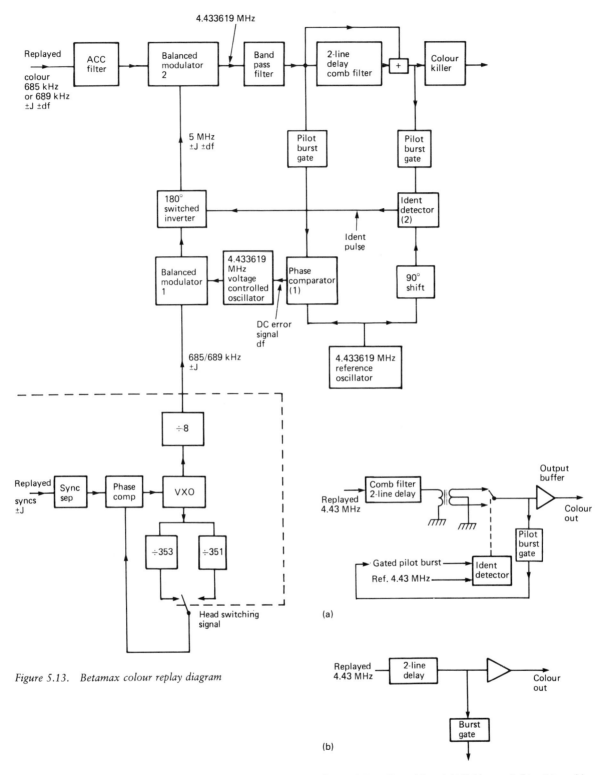

Figure 5.13. *Betamax colour replay diagram*

Figure 5.14. *Repositioned 180° ident switching (a), and later variations, VHS (b)*

Betamax colour replay system

The Betamax colour replay system is shown in block diagram form in Figure 5.13. Again the base carrier phase lock loop is similar to that of record mode with the exception that line syncs are from the replayed luminance signal. As in previous systems we have looked at, the phase lock loop is locked to replayed line syncs so compensation for changes in head to tape speed is present. This is again represented by jitter components J, and any jitter component present in the replayed colour signal will be cancelled by the jitter component in the replayed syncs. The jitter component J will appear out of the phase lock loop and be added to the intermediate frequency in balanced modulator 1. Subtraction of the J component will be carried out in balanced modulator 2.

As in the VHS system, Betamax has two phase comparators. One is identical to VHS in its function to correct for errors within 180°. Gated pilot burst is compared to a reference carrier of 4.433 MHz in phase comparator 1 and the DC error voltage controls a 4.433 MHz voltage controlled oscillator. The output from the voltage controlled oscillator is mixed with the base carriers on modulator 1 and fed via a switch to modulator 2, completing a correction loop for df within $\pm 90°$.

The second phase comparator is called an ident detector, although the name is different from VHS, the action is the same. Any errors that are greater than 180° when compared to the reference carrier in the ident detector will result in an ident pulse to the 180° switched inverter to invert the intermediate carrier thus bringing the error to less than 180° and within the range of phase comparator 1.

Note that the 180° switch for VHS is within the output of the phase lock loop and inverts the base carrier, whereas in Betamax the inverter is in the path of the intermediate carrier but the overall effect is the same.

Later variations

In later versions of the Betamax system, that is post model SL8000, the position of the ident switching system has been interposed between the output of the two line delay comb filter and the output buffer amplifier (see Figure 5.14a). 180° switching is achieved by a wide band centre tapped transformer, the switch being driven by the ident detector pulse from the phase comparator.

In later versions of VHS (HR7700/3V23) the burst gate has been moved (Figure 5.14b) and the gated burst is taken from the output of the comb filter. This brings the comb filter within the phase compensation loop thus improving the stability of the phase of the replayed colour signal.

The V2000 colour system

The V2000 system is similar to VHS and Betamax in so much as the replayed crosstalk is cancelled in a two line delay comb filter. In order to achieve crosstalk cancellation it is necessary that when a video head is replaying its track the colour crosstalk is picked up from the adjacent track in the familiar pattern of inverted line pairs (see VHS).

The colour undercarrier frequency for the V2000 system is 625 kHz and the colour undercarrier is 'processed' in record and replay in order to obtain crosstalk cancellation. In previous systems the cancellation was based on a 90° phase shift at line rate, but in the V2000 system the recorded colour signal is 180° phase inverted every *four* lines. The colour undercarrier is therefore recorded with four lines in normal (+) phase and four lines in inverted (−) phase, a total switching period of eight lines. A switching signal to provide for the subcarrier inversion is derived from f'_h, line pulses extracted from within the AFC phase lock loop and divided by 8.

This switching signal is referred to as the $\frac{1}{8}f_h$ signal and it is present throughout the recording process as a continuous signal. It is not confined to one video head or the other, but applied to both. Adjacent track colour crosstalk, as previously shown, has to be replayed in a vector pattern of inverted line pairs in order to be cancelled in a two line delay comb filter. These inverted line pairs in the V2000 system are derived from a combination of both the pattern of the vectors laid down on the magnetic track during recording and re-inversion of the four line inverted phase switching during replay.

In Figure 5.15 the recorded magnetic track pattern is shown, where the (+) and (−) symbols represent the burst, and hence the colour phase. In a 625 line video system, a $\frac{1}{8}f_h$ switching signal will complete 78 full periods with one line remaining.

$$625 \div 8 = 78 \text{ r } 1. \quad (78 \times 8 = 624)$$

The remainder of 1 is a single line offset between the switching signal and the video signal. It means that the switching signal will start one line earlier each subsequent frame and it will take eight

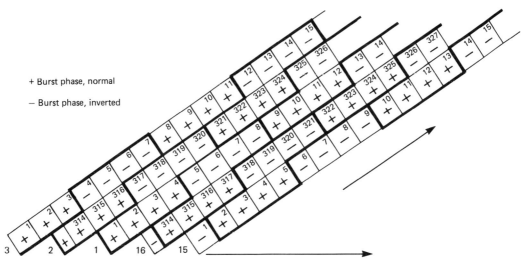

Figure 5.15. V2000 track patterning illustrating burst phase two line offset per track due to addition of 1½ line track offset and one line per frame shift

frames before repeating the sequence. Remember that each TV frame consists of two fields. Also each magnetic track laid down on the tape is one field. It can then be argued that the eight line switching signal, starting one line earlier for each frame, must therefore start half a line earlier for each field or magnetic track. This half line advancing shift between the $\frac{1}{8}f_h$ switching and the magnetic tracks is called a half-line offset, for each magnetic track. The recorded field tracks in the V2000 system are laid down side by side with a 1½ line offset between the starting points of each magnetic track (see N1700 system). We now have an accumulation of two offsets. A 'mechanical' offset between magnetic tracks, determined by tape speed and mechanical track angle, of 1½ lines, and an electronic offset of half a line due to the division of 312½ lines in one field by 8 (312½ ÷ 8 = 39 r ½). The result is a two line offset, the $\frac{1}{8}f_h$ switching between magnetic tracks. This offset can be seen in Figure 5.15, and the $\frac{1}{8}f_h$ switching signal advances two lines on each subsequent magnetic track. We now have the colour subcarrier at 625 kHz recorded for four lines positive phase and four lines inverted phase and offset by two lines between adjacent field tracks. The colour phase recorded onto the tape is as per Figure 5.15.

We now select a track for replay, say for example track 1. In Figure 5.16a, the diagram shows a sample of the magnetic track as recorded

with track 1 in the centre and tracks 2 and 16 either side.

When track 1 is replayed and re-inverted back to normal phase, any crosstalk pick up from either track 2 or track 16 is also replayed and inverted by the same switching signal $\frac{1}{8}f_h$. The colour replay phases are shown in Figure 5.16b. It can be seen that the colour phases for track 1 are all re-inverted to normal positive phase. However the replayed crosstalk from track 2 or 16 reveals the required replay crosstalk pattern for cancellation – inverted line pairs!

Figure 5.16c illustrates the crosstalk cancellation in the two line delay line comb filter. If we take the replay of line 5 of track 1 along with lines 319 (track 2) and 316 (track 16) and apply it to the delay line and the sum network, coming out of the delay line will be line 3 of track 1 along with crosstalk from lines 317 (track 2) and 314 (track 16). In the sum network replay lines 5 and 3 will add together whereas lines 317 and 319, 316 and 314, will cancel.

The V2000 system acts the same way as others to cancel the crosstalk in a two line delay comb filter but the crosstalk pattern is derived in a different way.

The system does of course rely on the replay $\frac{1}{8}f_h$ switching signal being synchronised with the record $\frac{1}{8}f_h$ signal in order that the video track being replayed is re-inverted correctly and so a method to achieve this has been devised.

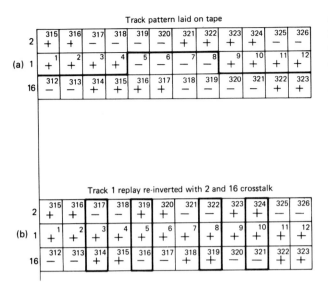

Track pattern laid on tape

	315	316	317	318	319	320	321	322	323	324	325	326
2	+	+	−	−	−	−	+	+	+	+	−	−

	1	2	3	4	5	6	7	8	9	10	11	12
(a) 1	+	+	+	+	−	−	−	−	+	+	+	+

	312	313	314	315	316	317	318	319	320	321	322	323
16	−	−	+	+	+	+	−	−	−	−	+	+

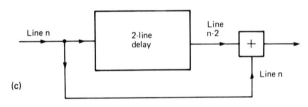

Track 1 replay re-inverted with 2 and 16 crosstalk

	315	316	317	318	319	320	321	322	323	324	325	326
2	+	+	−	−	+	+	−	−	+	+	−	−

	1	2	3	4	5	6	7	8	9	10	11	12
(b) 1	+	+	+	+	+	+	+	+	+	+	+	+

	312	313	314	315	316	317	318	319	320	321	322	323
16	−	−	+	+	−	−	+	+	−	−	+	+

(c) Line n → 2-line delay → Line n-2 → + (adder); Line n also fed to adder.

Example:	Replay (n)	Phase	(n-2)	Phase	Output of adder
Replay line	5	+	3	+	+
Track 2 Xtlk	319	+	317	−	0
Track 16 Xtlk	316	−	314	+	0
Line	10	+	8	+	+
Track 2 Xtlk	324	+	322	−	0
Track 16 Xtlk	321	−	319	+	0

Figure 5.16.

Practical application of V2000 colour record/replay

Record

Figure 5.17 shows the block diagram of a Grundig 2 × 4 VCR colour circuit. Some of the signals are multipath and therefore require record/play switching, done by switching power supply rails. In the record mode the video input signal is filtered by a 4.43 MHz filter and then passed to an ACC level control circuit.

Input 4.43 MHz is split up into two paths, one is the main recording path into balanced modulator 2 for down conversion to 625 kHz. It then passes via S1 and a low pass filter out to the recording drive circuitry where it is mixed with the luminance FM carrier. A second path is for ACC level control and to derive a 4.43 kHz conversion carrier. The 4.43 MHz colour signal bypasses the balanced modulator 2 via switch S2. From S2 the 4.43 MHz signal passes through the comb filter, not for any crosstalk cancellation at this stage. A bandpass filter and the comb filter are used as a convenient filter system. Out of the delay line this signal is sent to an ACC and APC integrated circuit, similar to one used in colour TVs.

In record mode the IC has three functions:

The burst is gated and used for ACC level comparison and a control signal is sent out to the ACC circuit in the input path. It is interesting to note that a large portion of the input path is within the ACC loop avoiding circuit tolerances affecting the colour under record path.

An automatic phase control loop (APC) compensates for transmitted path phase errors and provides a controlled 4.43 MHz carrier to balanced modulator 1. Compensation is carried by the 5.05 MHz intermediate carrier to balanced modulator 2.

Lastly the ACC level operates the colour killer; note that the record colour killer amplifier is in the record signal output path. This prevents spurious colour signals from being recorded.

Input video is also fed via a line sync separator to the AFC loop. A 625 kHz voltage controlled oscillator is divided by 40 to 15.625 kHz or derived line syncs f'_h. f'_h is then compared to incoming line syncs, f_h, in a phase comparator, an error voltage being used to control the VXO, thus completing the phase locked loop. 625 kHz from the VXO is the base carrier, phase locked to line syncs and carrying any large transmitter path phase error correction. Balanced modulator 1 mixes the base carrier at 625 kHz with 4.43 MHz from the APC loop to give the intermediate carrier at 5.05 MHz.

The colour undercarrier switching signal $\tfrac{1}{8}f_h$ is derived by dividing f'_h from the AFC loop by eight. It is fed via an inverter switch, which does not operate in record, to an intermediate carrier switch.

A centre tapped tuned transformer provides 5.05 MHz in both positive and negative phases at its two end outputs. These outputs are selected alternatively by the $\tfrac{1}{8}f_h$ switching signal to give four lines in normal phase and four lines in inverted phase. The input colour carrier is both down converted and inverted every four lines in balanced modulator 2.

103

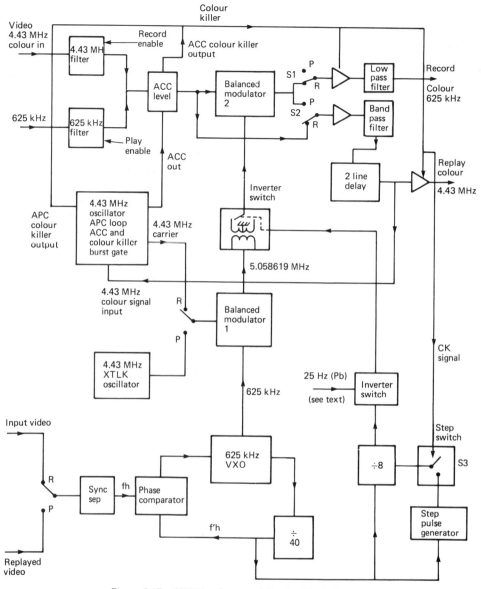

Figure 5.17. V2000 colour record/replay block diagram

Replay

The replay path of the 625 kHz colour undercarrier is to balanced modulator 2 via a 625 kHz low pass filter and the ACC level control circuit which was used in record. In balanced modulator 2 the replay signal is upconverted back to 4.43 MHz which is in turn passed to the two line delay comb filter via S2 and a bandpass filter.

Out of the delay line the replayed 4.43 MHz is taken through an output buffer amplifier, which also has the replay colour killer operating upon it.

The replay colour signal is also taken back to the ACC and APC integrated circuit. Within the IC the replayed burst is used for ACC level detection; again note that both balanced modulator 2 and the comb filter are within the ACC loop. Replayed burst is also phase compared to an internal oscillator within the APC loop; if the replayed colour signal is not in phase then the APC loop will detect this and operate the colour killer.

As in previous systems, replayed video is fed to the AFC loop via a line sync separator for replayed

104

colour undercarrier frequency correction. As the 2 × 4 video recorders have slow motion, still frame and multiple speed facilities, then the replayed signals will be subject to frequency variations. Any changes in head to tape speed will alter the replayed colour undercarrier and line syncs.

Changes in line sync speed will change the 625 kHz VXO frequency because of the AFC phase locked loop action. A 4.43 MHz crystal oscillator provides a reference for the replayed signal, being used to upconvert 625 kHz base carrier to 5.08 MHz intermediate carrier. Any speed corrections will also upconvert and correct the replayed colour undercarrier in balanced modulator 2.

The replay circuit also has to correct for any record to replay phase errors. In order for crosstalk cancellation as previously described to be effective, then the four line inversion applied during record has to be re-inverted back to normal. This requires that the $\frac{1}{8}f_h$ switching signal during replay correction *has* to be synchronised to one applied during record. A system has been devised that will detect the fact that the replay $\frac{1}{8}f_h$ is out of synchronisation and correct it. If the replay $\frac{1}{8}f_h$ is out of step then phase errors occur and the colour killer operates. In replay $\frac{1}{8}f_h$ is derived from f'_h as in record. The reason for using f'_h and not f_h is that any dropouts might affect f_h resulting in a lost pulse and causing replay $\frac{1}{8}f_h$ to drop out of synchronisation. f'_h will not be affected by the loss of one or two replayed line pulses, since time constants in the AFC loop will carry the oscillator over any sync losses in a similar manner to a TV flywheel sync circuit.

In the situation where the replayed $\frac{1}{8}f_h$ is not synchronised and the colour killer is operated then stepper switch S3 is also closed. A step pulse generator provides extra line pulses to the divide by eight counter and it accelerates the count. Replay $\frac{1}{8}f_h$ is therefore advanced by one line at a time; when it matches up to the recorded switching signal there is no longer a replayed colour signal phase error, and the colour killer is removed. So are the stepper pulses.

This system ensures that the four lines of colour undercarrier which were inverted during record are correctly re-inverted in play by synchronising $\frac{1}{8}f_h$ between record and play. This method eliminates the need for balanced modulator 1 to have an upconversion carrier at 4.43 MHz, which is derived from a VXO and phase detector.

There is a small aside to this particular system, to be found in the inverter switch in the $\frac{1}{8}f_h$ signal path. It is driven by 25 Hz pulses from the drum servo. The reason for the need of this 25 Hz inverter is to be found in the video head replay pre-amp. The two video heads are connected to a centre tapped transformer in replay. Ch.1 is connected to the 'in phase' output, however, Ch.2 video head's replay signal is connected to the other end and is subject to 180° phase inversion. Whilst this replay signal inversion does not affect the luminance FM carrier, it does affect the replayed colour undercarrier which is replayed in the wrong phase. In consequence, every time Ch.2 video is replaying, the $\frac{1}{8}f_h$ switching signal is inverted to compensate by using the drum servo flip-flop or (HI) signal to drive the in line $\frac{1}{8}f_h$ inverter.

A number of operational points are worth noting, although actual circuit practice varies from model to model.

The colour killer is a schmitt trigger circuit driven from the ACC voltage level within an IC. Also the IC is fed from an output from the phase detector, if the phase detector is operating rapidly, as it does in the case of multiple phase errors. Its output switching is integrated and fed to the ACC circuit to operate the colour killer.

The ACC correction voltage is clamped at the start of each field, driven by the HI pulses from the servo.

Later version of VHS AFC loop

Referring back to Figure 5.8 (VHS colour record system) the original designs were based on the use of a phase locked loop producing an output of 625 KHz. This frequency was chosen because it is easily divided by 40 to line frequency fh (15.625 KHz) permitting a simple design of AFC loop. However, up conversion from 625 KHz to 5.06 MHz in balanced modulator 1 has to be done by using a non-standard crystal frequency of 4.435571 MHz. No such problem arises in Betamax machines; a standard 4.433619 MHz crystal oscillator feeds the up convertor (Figure 5.11) because the output frequency of the AFC loop is the same as that of the recording colour under carrier. Later versions of VHS machines (e.g. Sharp, Grundig VS200) can be seen to use a standard 4.433619 MHz crystal in the sub convertor stage. Here the AFC loop is arranged as shown in Figure 5.18.

The voltage controlled oscillator on the right runs at a frequency of 5.015625 MHz (321 × fh).

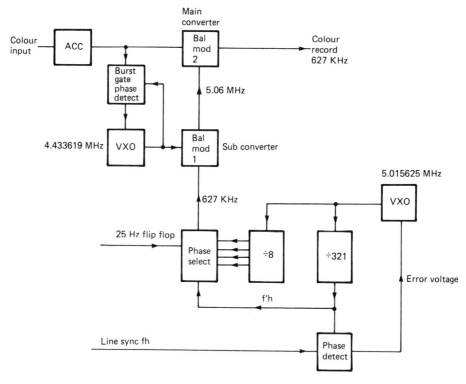

Figure 5.18.   627 Khz AFC loop

The phase locked loop is completed by a ÷321 counter whose output (f'h) is compared to fh in a phase detector; the latter's error voltage output steers VXO frequency to maintain lock. An output is taken from the VXO and divided by 8 to obtain the base frequency of 626.953 KHz. This is identical to the colour under carrier frequency, and is related to line frequency by a factor of 40.125: 626.953 KHz = 15.625 KHz(fh) × 40.125. Alternatively, 626.953 KHz = (40 + ⅛)fh. As in the Betamax system, this approach allows the balanced modulator 1 to be fed with 4.433619 MHz as an up converting carrier. The advantage is that the carrier can now be phase locked to incoming colour burst, and the APC discriminator used as a source of a record colour, killer control voltage.

During replay the 4.433 MHz oscillator free-runs; replay colour phase is governed by another combined APC and AFC loop, locking playback colour to a separate 4.43 MHz oscillator. It is possible to use the same oscillator for both purposes and this is done in some models. Figure 5.9 shows how replay colour correction is carried

out by two loops. Frequency correction is achieved by the AFC phase locked loop and the generation of a base carrier, while phase correction is made by the use of two phase detectors and a VXO into balanced modulator 1. A point worth noting here is that some later video recorder circuits refer to balanced modulator 1 as the sub converter and balanced modulator 2 as the main converter – our diagrams here have been labelled accordingly.

A refinement of the technique illustrated in Figure 5.9 is shown in the diagrams of Figure 5.19 (record) and 5.20 (replay). In the record mode there is little change to the down conversion system; a 4.433619 MHz voltage controlled oscillator (VXO) is slaved to incoming gated burst signals and checked against them by a phase detector to provide colour killer action in the event of weak or absent burst. Incoming line sync locks the AFC loop (bottom of Figure 5.19) in the normal way to derive the phase locked carrier at 625 KHz. Down conversion is carried out in the traditional way, but the local reference (sub-converter feed) comes from an autonomous crystal oscillator at 4.435572 MHz rather than the VXO

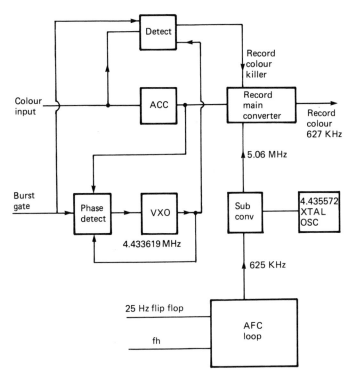

Figure 5.19.   Late version of VHS colour record system

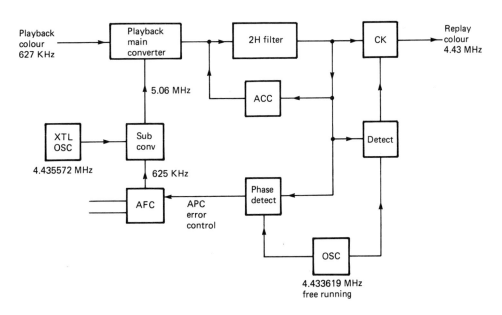

Figure 5.20.   VHS up-conversion and de-jittering, late models

of previous systems – the significance of this will be seen in a moment when we study the replay system.

The replay block diagram of Figure 5.20 shows the main changes brought about by the use of digital techniques and large scale integration. The most striking feature is the presence of an autonomous crystal oscillator driving the sub converter, where before there was a VXO being 'steered' for phase correction purposes. Here phase correction is carried out by comparing replayed burst with the output of a stable crystal-based 4.433619 MHz oscillator – the same one as was used during record, but now free-running. The resulting error signal is *added* to the error signal in the AFC loop, permitting both phase and frequency correction to be carried out via the 625 KHz base carrier link between the AFC and sub converter blocks. Two phase detectors are used as before, one giving $\pm 90°$ correction and the

other for errors greater than 180°; their operational principles are described in Figure 5.10.

Advanced AFC/APC loop

The AFC/APC control system is shown in Figure 5.21; it is a complex digital operation ensuring a high degree of accurate control. The starting point for replay AFC control in a phase locked loop is off tape line sync. Instead of the conventional phase detector a pulse counting system is used. Replayed line sync (fh) is divided by 8 in a counter which has two outputs; a symmetrical squarewave with a period of 8 h; and a single pulse at *intervals* of 8 h. The negative section of the symmetrical squarewave closes gate 1 which passes the output of a 160 fh (2.5 MHz) VXO to a $\div 160$ counter. Since the gate is closed for a period of 4 h the number of pulses to the counter is $4 \times 160 = 640$. A following counter divides by 4, and it has two

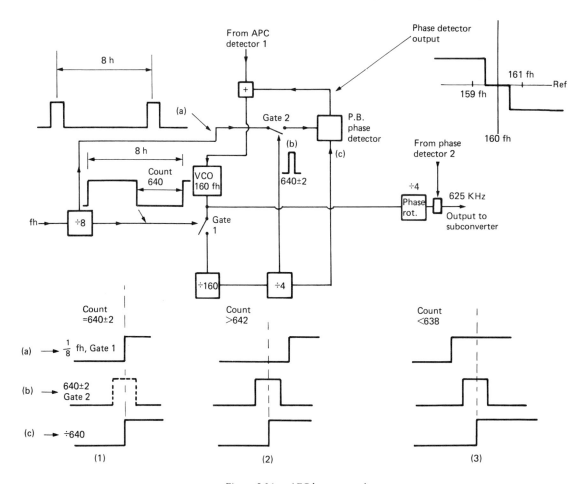

Figure 5.21.   AFC loop control

outputs: one to the phase detector at a frequency of fh (or f'h); and one to gate 2. The output to gate 2 is rather special – it consists of a short pulse, and is only present if the count via gate 1 is less than 638 or greater than 642; for counts between these two values no output pulse is produced and gate 2 stays off. For other count values gate 2 comes on, permitting comparison of waveforms a and c in the phase detector.

Operation

If the replayed line sync rises slightly in frequency due to an increase in head to tape speed the ⅛fh square wave to gate 1 becomes shorter in period. Less than 638 pulses are counted and timing diagram 3 is relevant. Because the count is outside the range 640 ± 2 pulse (b) appears to turn on gate 2, passing waveform (a) into the phase detector. Waveform (c) is a latch driver such that the level of waveform (a) is latched when (c) goes high. Since (a) is high at this time the phase detector output goes high, increasing VXO frequency so that the number of counted pulses during the 8 h period increases.

Conversely, if the replayed line sync frequency should fall due to a decrease in head to tape speed the period of the ⅛fh square wave becomes longer. More than 642 pulses are counted and timing diagram 2 becomes relevant. Again pulse (b) will be generated (count outside range of 640±2 pulses) so waveform (a) is gated to the phase detector. This time (a) is low when (c) latches the phase detector, whose output will go low as a result. The 'low' pulls down VXO frequency to restore the count to 640.

Once the VXO becomes 'locked' to replayed line syncs and no frequency correction is required, the pulse count via gate 1 will be between 638 and 642. This satisfies the 640±2 requirement in the ÷4 counter, so waveform (b) is not produced, see timing diagram 1. Gate 2 stays off, then, and no (a) pulses are present in the phase detector when it is latched by waveform (c). The output from the phase detector to the VXO stays therefore at a fixed reference level – it is upon this level that APC control is superimposed.

When the count is 640±2 the VXO must be running at a frequency of 160 fh±0.5 fh and the playback AFC detector produces a reference output. The figure of 0.5 fh is equivalent to 180°. Correction within 1 cycle, once the AFC loop is locked, is the responsibility of the APC loop, in which two phase detectors are used. One provides 'analogue' correction in the form of a variable voltage to the 160 fh VXO, and operates within the range +90° to −90°. If the error becomes large enough to fall outside this range, the second (phase detector 2) switches in an invertor in the 625 KHz output path to effectively invert the phase by 180°.

Large scale integration of colour processing

Colour record

Composite video signal enters the IC on pin 2 where it is buffered and selected for record and replay and then exits to a band pass filter between pins 31 and 29 where the colour component is filtered out. In pin 29 is the automatic colour control or ACC to stabilise the colour signal from external variations. The output of the ACC is sent to the record main converter but it is also routed back to the ACC detector where the burst is gated out and compared to a reference. Any deviation in level is compensated for by altering the gain of the ACC amplifier, the error signal is damped by the capacitor on pin 30 against erroneous instantaneous changes in the burst and colour level.

Within the record main converter the colour signal is down converted to 627 kHz and comes out of pin 5 to the colour record amplifier via the low pass filter.

Record automatic phase control

Within the IC the colour signal is routed via an APC switch to the record APC detector on input A. The burst is gated out by a gating pulse on input C whilst a 4.443618 MHz reference carrier is on input B. If the burst is absent or the wrong frequency the APC circuit detects this and operates the record colour killer section of the record down converter. A fairly common fault can cause the symptom of no colour record in this area and it can be traced usually to the crystal reference frequency being incorrect often due to the crystal being faulty. The colour carrier is buffered out on pin 13 and is frequently used in the servo sections as a reference frequency for phase control.

Automatic frequency control

You will be aware from this chapter that for accurate frequency convertion of the colour signal

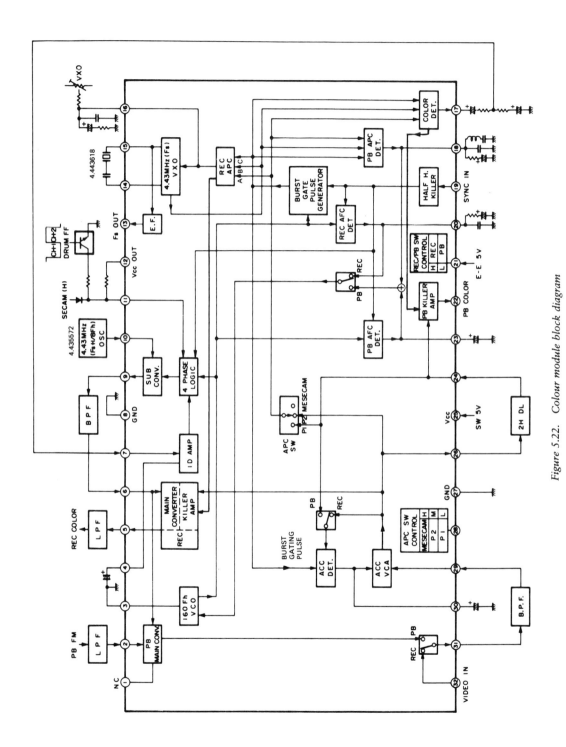

Figure 5.22. Colour module block diagram

110

the line syncs (fh) are utilised, these enter this IC upon pin 19. The main voltage controlled oscillator which runs at 160 fh is attached to pin 3 and is counted down to 625 kHz with the (-90°/line) phase shift in the 4 phase logic block and fed to the subconverter. This 160 fh signal is also counted down to fh/2 where it is compared to fh/2, out of the half H killer block from pin 19, in the record APC block. The record error signal sent back to the 160 fh VCO via a record selector switch. The reason for the record APC working at half line is to accommodate the two line repetition rate of a PAL signal.

Subconverter

The derived 625 kHz is mixed with 4.435572 MHz (Fs + 1/8 fh) to produce a 5 MHz convertion carrier which comes out of pin 9 and via a bandpass filter to pin 6.

Replay

Replayed FM carrier is presented to pin 2 via low pass filter to the playback upconverter block to produce 4.433 MHz colour carrier out of pin 1. The replayed colour signal also is subject to automatic colour control and follows a similar path to the recording signal. The main difference is that the 2H delay line comb filter is within the ACC path between pins 26 and 24. Playback colour comes out of the IC on pin 22 via the playback colour killer.

Playback AFC and APC

Playback AFC is similar to that of record with playback syncs (fh) on pin 19 controlling the phase lock of the 160 fh VCO with the error signal switched over from record to play. The main difference is the addition of the APC loop for phase correction. The gated replay colour burst and the 4.43 MHz reference are compared in phase in the PB APC block, the phase error output is added to the error signal from the AFC loop.

In addition to the APC detector there is the colour detector which has an output on pin 17. If the colour signal is more than 180° out of phase the colour detector sends an error voltage to the ident amplifier from pin 17 to pin 7. Upon receipt of the error signal the ident amplifier compensates for the error by inverting the phase of the current output of the four-phase logic circuit. Once the error is within 180° the playback APC detector output will bring the phase of the replayed colour signal into line by a linear correction voltage added to the AFC voltage back to the VCO. If the replay APC corrections to the colour detector are very frequent, indicating a problem, then it activates the colour killer in the playback amplifier on pin 22.

The APC loop is switchable and is normally found in the second position P2. In P1 the 2H delay line is brought within the APC correction range. In SECAM the APC loop is inhibited as it has no use in this colour system.

6

SYSTEMS CONTROL

The function of the systems control circuits within a videorecorder is to organise the handling of the videotape in order to minimise damage. This is achieved by monitoring the operating modes and controlling the operation of the mechanical section in an orderly manner. In early models, with mechanical keys, this is partly done with mechanical interlocking levers operated by solenoids. In later models control is under a microprocessor programme, operating solenoids. Levers and interlocks tended to be replaced by operational motors. Basic models had logic ICs and transistor interfaces to drive the solenoids. Expansion of the systems capabilities brought about the appearance of large scale integrated circuits which contained the logic ICs in one package.

The advent of front loading machines with touch button controls meant that more sophistication was required and the microprocessor took over control functions. The systems control is abbreviated by some manufacturers to 'Syscon' whilst others abbreviated 'Mechanism Control' to Mechacon.

General operation

Until a cassette is present, the systems control ensures that all operations are inhibited; in the case of VHS this is also the situation if the lamp in the cassette compartment fails.

The end of tape sensors are optical and consequently the cassette would be damaged if the lamp failed and tape transport was not disabled. If the lamp fails during play, the control system enters Stop and then Unthreading, finally into Stop once more. The tape is also unthreaded if the end of tape sensor is activated in either Play or Record.

End of tape detection in either Fast Forward or Rewind will result in stop mode being entered. All forward functions are inhibited when the end of the tape is reached and the end sensor is activated. Only Rewind can be selected, but in more automated recorders the rewind function is automatic. At some point during the Rewind mode the tape

start sensor is activated, this initiates Stop and further rewind is inhibited, and only forward functions can be selected.

Counter Search operation will cause the systems control to enter Stop when the counter passes 0000.

The system control resets all operations whenever power is applied, although if power is removed during Play the tape remains threaded. Unthreading takes place when power is re-applied and the circuit resets to the Stop mode.

Tape protection control

The systems control circuits monitor the tape transport mechanics and serve to protect the tape, as far as possible, from damage. The take up spool rotation is monitored and if it ceases to rotate during Play then unthreading takes place and Stop is entered. There is a short time delay of 4–5 seconds to prevent erratic operation due to a momentary pause during normal working rotation. In basic models only the take up spool is monitored; in the sophisticated models the supply spool is also monitored. If the tape does not travel after an operational function is selected, again monitored by the take up spool, within 5 seconds, Stop mode is entered.

The video head drum is also monitored, using the derived flip-flop signal, and if rotation ceases in Play or Record then unthreading takes place after 5–6 seconds. In some models the pinch wheel is retracted whilst the head drum is stalled and unthreading will not take place until rotation is restored or the systems control circuits are reset by switching the machine Off and then On again.

Some videorecorders have a slow motion or still frame facility, which requires that the supply and take up spools stop rotating in a Play mode but without Stop or unthreading being initiated.

This is the 'Play Pause' or even 'Record Pause' mode of operation. The system control circuits inhibit unthreading during still frame and also start a time elapse counter. This counter or timer is to prevent tape damage by limiting the length of time that the stationary tape can be wrapped around the rotating video head drum. Timer periods can vary from 6 minutes to 14 minutes depending on the model and manufacturer.

Figure 6.1 is a practical systems control circuit diagram. The logic gates used in this circuit are NAND gates. The operation of a NAND gate is as follows: the output of a NAND gate is normally high when it is in the 'off' condition. It requires *both* inputs to go 'high' before the gate switches on and the output changes state to 'low' or zero volts.

In the Stop mode the following conditions are present. Play switch S2, and OPE switch, S4, are in the NC position; there is no OPE 12 V therefore transistor X19 is off. As X19 is off, MDA control transistors are on shunting the motor drive voltages to ground, so the motors will not rotate. When the Play key is pressed S2 and S4 are put into the NO position.

With a cassette inserted, switch S12 is put into the NO position, X1 in the end sensor circuit is switched on and reverse biases diodes D3 and D4. The gates IC3A, B and C are all off because the Rewind key is not pressed, the outputs of these gates are therefore high. Gate IC3D is also off as the End sensor is off and pin 13 of IC3D is low, the output of IC3D is also high. All three inputs to gate IC2B are high, being driven from IC3, so this gate is on and its output is low. IC2C is an inverter, which is off and its output is high.

Switch S4 provides the OPE 12 V power supply. Transistor X2 is turned on and reverse bias diode D2 which has charged C1 to 12 V; C1 cannot discharge quickly. IC4A is on as all three inputs are high, its output is low so X4 is held off enabling the 12 V OPE supply to turn on X19. X19 switches off X20 and X21 and the motors rotate.

When threading takes place, switch S7 is in the NO position so the transistor X1 is off and with the Play switch pressed S2 is in the NO position. Gate IC1B is turned on with both inputs high, the low output holds IC2A off. IC2A output is high which via D1 and R9 maintain C1 charged during threading. At the end of threading S7 changes to the NC position and IC1B turns off.

In order to maintain rotation of the motors after loading C1 must be kept charged to a high level to prevent IC4A from turning off. This is achieved by combining the flip-flop signal at 25 Hz, which is derived from the rotating video head drum in the audio/servo PCB, with a 1 Hz pulse derived from the rotating take up spool. The two rotational

signals are combined in IC1D and the signal on pin 11 is as in Figure 6.2; the signal is inverted through IC2A. The high portions of the signal top up the charge on C1; in the low periods C1 discharges, at a very slow rate, but not enough to switch off gate IC4A. This is due to the discharge path being only through R8 which is 1 Mohm.

Safety function

If the video head drum ceases to rotate, for whatever reason, the flip-flop signal is no longer present. In this case pin 8 of IC1C remains high pin 9 is also high so IC1C switches on, and its output goes low. IC1D then switches off, cutting off any signals to IC2A, which switches on due to all three inputs being high, C1 discharges below the 6 V input threshold of IC4A pin 12 and IC4A switches off; X4 switches on to cut off motor power. IC4A operates the stop solenoid via IC4B and C which releases all of the function keys. A similar train of events occurs if the take up spool stops rotating and the 1 Hz switching signal is no longer present.

Transistor X3 remains turned on and the low voltage present on its collector also turns off IC1D causing the stop solenoid to operate and the motors to switch off. At this point in time, S2 and S4 are reset to the NC position but the tape is still in the threaded up position.

The unloading switch S7 remains closed and 12 V travels via diode D7 to transistor X19 to hold it on, maintaining rotation of the motors whilst unthreading takes place. When unthreading is completed, S7 returns to the NO position, X20 and X21 then switch on to stop the motors.

If the End Sensor is activated, when the video cassette has played its full length, the End Sensor turns on. IC3D turns on as both inputs are then high, its low output turns off IC2B which turns on IC2C. The resultant low output turns off IC4A, which as previously described, turns the motors off and operates the Stop solenoid.

In this condition, no other function will operate other than Rewind. The Rewind Key will turn on IC3A which in turn will turn off IC3D; this will allow IC2B to turn on again, turning IC2C off and allowing IC4A to turn on again and also the motors. If the start sensor is activated, or the counter search switch operates, IC2B will be turned off again causing IC4A to turn off and hence stop the motors. Note that if the start sensor is activated, then IC3B, which turns off IC2B, will prevent the Rewind operation.

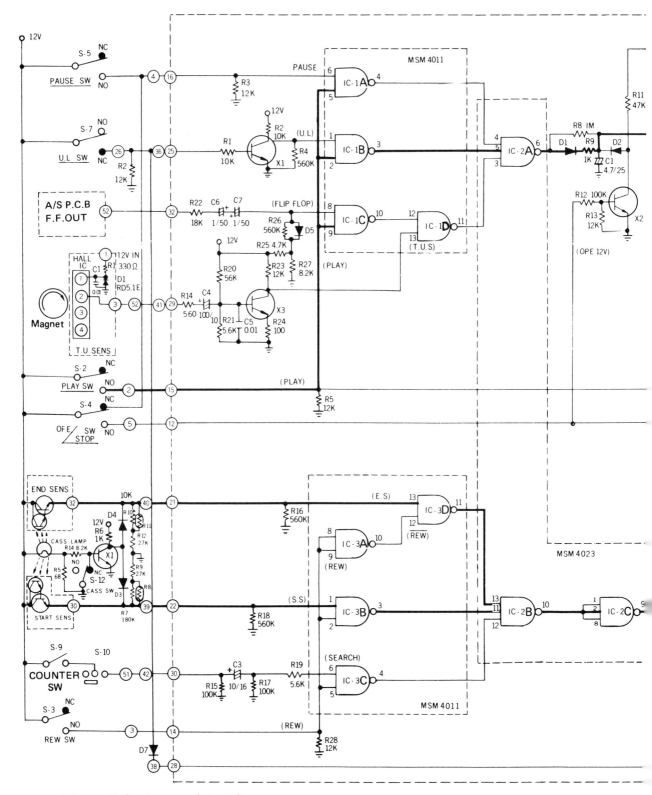

Figure 6.1. Practical systems control circuit diagram

114

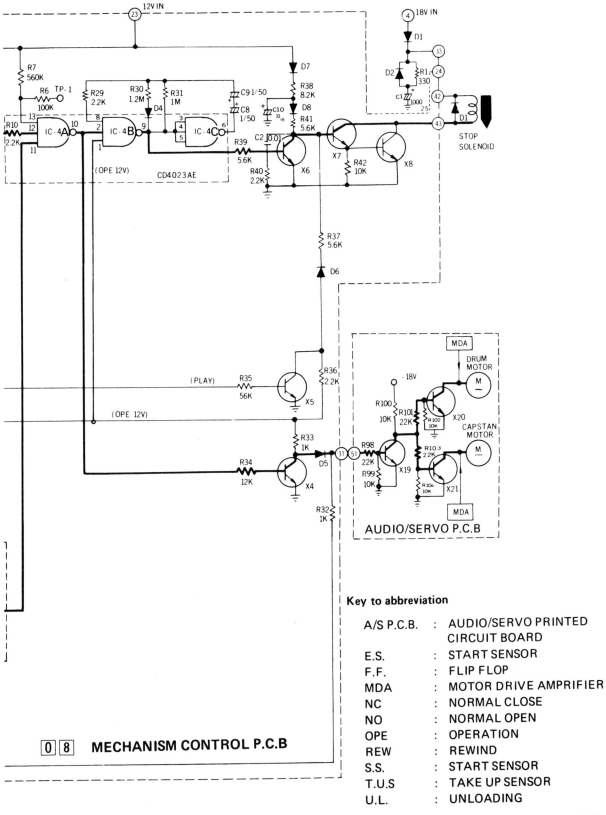

Key to abbreviation

A/S P.C.B.	:	AUDIO/SERVO PRINTED CIRCUIT BOARD
E.S.	:	START SENSOR
F.F.	:	FLIP FLOP
MDA	:	MOTOR DRIVE AMPRIFIER
NC	:	NORMAL CLOSE
NO	:	NORMAL OPEN
OPE	:	OPERATION
REW	:	REWIND
S.S.	:	START SENSOR
T.U.S	:	TAKE UP SENSOR
U.L.	:	UNLOADING

08 **MECHANISM CONTROL P.C.B**

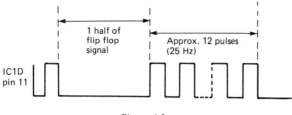

Figure 6.2.

Rewind is allowed if the counter search operates as this is in pulse form to operate the stop solenoid via IC3C, IC2B or C and then IC4A, B and C.

Pause mode

In the pause mode it is necessary to maintain the motors in the operate condition whilst the Take up spool has ceased rotation, by inhibiting the safety circuit. The main pause operation is a mechanical leverage system, but our systems control has to take care of the logic.

The pause switch S5 is put into the NO position, turning on IC1A. Pin 4 of IC1A goes low and turns off IC2A, the output of which goes high. The high output of IC2A holds C1 charged and IC4A on, preventing C1 from discharging in the absence of any of the pulse waveforms, and of course the motors still rotate. The mechanical section of the pause key lifts off the pinch wheel drive and brakes the tape spools.

More sophisticated systems control can be fitted into video recorders and in general the first ones to emerge were based on four-bit microprocessors. In these initial micro based systems control circuits the micros are standard four-bit ICs which were mask programmed during manufacture.

For example, the JVC HR7700 used a four-bit micro that has previously been used in automatic washing machines, only the inbuilt programme was different. As I used to say to students during HR7700 training lectures 'if a JVC HR7700 suddenly goes into spin/rinse then you will know why!' In the mechanical control or systems control of the HR7700/3V23 a large number of interface ICs and transistors can be found to decode the microprocessor outputs and arrange them into a drive more suitable for the solenoids and motors to be driven from the micro. Inputs in the form of data had also to be arranged into a form suitable for the micro by the use of data selectors. Data selectors are a method of strobing input data, up to 32 different inputs into only eight input ports;

later versions of data selectors have emerged as input expanders as you will soon see.

Later generations of systems control microprocessors are much more 'custom designed' for the job that they had to do. In this respect less peripheral discrete components are required to match the micro in/out ports to the pushbuttons and tape deck mechanical components. Only interfaces to motors are required to perform the functions of being able to reverse the motors and run them at various required speeds. The more functions that a video recorder has within its facilities then the more complex the system control circuit is.

Facilities have increased, from still picture, then cue and review, then the addition of back space editing and slow motion with noise free replay. Currently the peak of video recorder capabilities is insert editing with clean-in switching and clean-out switching.

The JVC HR7600/7650 systems control

An outline of the JVC HR7600 mechacon (mechanical control) system is shown in Figure 6.3.

Functions from the infra red remote control or from the front panel pushbuttons are fed to a series-to-parallel converter. These are the Stop, Play, FF, Rew, Pause, Record etc., and are usually called the Mechacon Functions. Each individual function is assigned as 7-bit binary code, preceded by a 3-bit 'key code'; in binary it is 100. This key code is to tell the video recorder that it is a JVC HR7650 remote control that is generating the binary information and not a Sony or other machine, that is to say, it is an identification code.

A 10-bit binary word is issued by either the IR remote control or by the front panel function pushbuttons, which are actually switching a series pulse encoder to produce the required binary word.

Serial-to-parallel converter

Series binary information is a train of pulses, but the microprocessor does not read series pulses; it wants them all at once, in parallel. Hence we have a series-to-parallel decoder which is really a shift register.

There are eight outputs from the serial-to-parallel decoder as seen in Figure 6.4. Seven of the

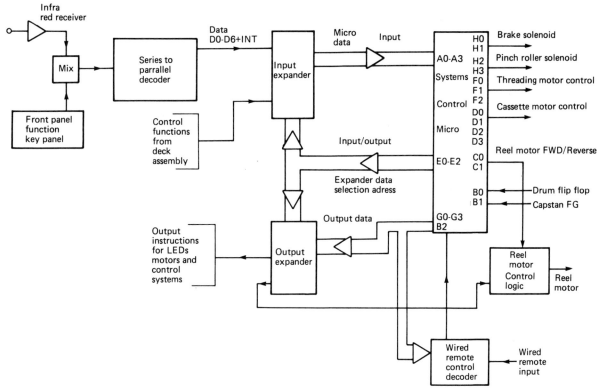

Figure 6.3. Mechacon system, JVC HR7600

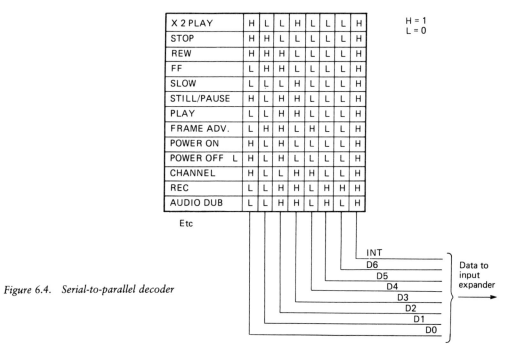

	D0	D1	D2	D3	D4	D5	D6	INT
X 2 PLAY	H	L	L	H	L	L	L	H
STOP	H	H	L	L	L	L	L	H
REW	H	H	H	L	L	L	L	H
FF	L	H	H	L	L	L	L	H
SLOW	L	L	L	H	L	L	L	H
STILL/PAUSE	H	L	H	H	L	L	L	H
PLAY	L	L	H	H	L	L	L	H
FRAME ADV.	L	H	H	L	H	L	L	H
POWER ON	H	L	H	L	L	L	L	H
POWER OFF	L	H	L	H	L	L	L	H
CHANNEL	H	L	L	H	H	L	L	H
REC	L	L	H	H	L	H	H	H
AUDIO DUB	L	L	H	H	L	H	L	H

Etc

H = 1
L = 0

INT
D6
D5
D4
D3
D2
D1
D0

Data to input expander

Figure 6.4. Serial-to-parallel decoder

117

outputs are the function binary information and are labelled D0–D6; the eighth is an interrupt output which goes high whenever function data is being received. This is used to tell the micro that it is receiving valid function binary data.

The data output from the serial-to-parallel converter does not go directly to the microprocessor as there is an input expander (or rather compressor) in the way.

Input expander

Input expanders are used to route information into the microprocessor under the programmed control of the micro. In this case the micro has only four input ports and there are 28 information data lines. Data lines are therefore split into four sets of seven. Each set of seven is allocated to one micro input port and each data line is switched into the micro port in turn.

Each of the four sections of the expander is similar to a mechanical seven way wafer switch, so the whole expander is similar to a large seven way switch with four wafers on the shaft giving a four

pole seven way switch with all of the four sections working as seven way selectors in parallel.

However, in a mechanical switch you are restricted to a sequence of 1, 2, 3, 4, 5, 6, 7. In this equivalent electronic four pole seven way switch no such restriction exists and the selectors, in parallel, can go from 1 to 4 to 7 to 3 to 2 or whatever is required. The control of the switching is done by the micro via three output ports E0, E1 and E2.

Three-bit binary ports can produce $2 \times 2 \times 2$, or eight possible binary combinations; however the expander only requires 7, so 000 or low, low, low, is not used.

In this way the micro can issue a three-bit binary word say, for instance, 110 (H, H, L). then the start sensor is switched to port A0, the End sensor is switched to port A1, cassette lamp to port A2, and the record safety switch to port A3. In the next instant the micro may produce 010 (L, H, L) and we will have D5 to A0, D4 to A1, D3 to A2 and D6 to A3. Consequently, acting under the control of its programme the microprocessor can select 4 out of 28 in four parallel combinations of 1 out of 7, selecting any four particular inputs at any instant in time.

Function data, 00–06, is selected in two stages, 010 and 011. The micro therefore feeds the information into a temporary store in order to read the information fully.

Let us look at the information which is available for selection:

We know that Data D0–D6 and Int is one of the thirteen possible mechanical functions.

Camera pause is for camera recording which will also put the recorder into back space editing (see Editing).

Start Sensor, Optical Sensors, which will be used to stop the tape at the completion of rewind.

Supply Reel FG Pulses generated by the supply reel rotating are integrated to maintain a low level, high when stopped.

Cassette Detect, Optical sensor, used to determine that a cassette is inserted into the cassette compartment. Low with a cassette in. (Note that this machine is a front loader and has a cassette compartment 'lift' mechanism.)

Timer Rec Start, signal from the timer circuit to put the system into record for a timer recording.

End Sensor, optical sensor to stop the tape at the end of the cassette in Play, Record or FF.

Insert Edit Switch, puts system control into insert Edit Mode (see Editing).

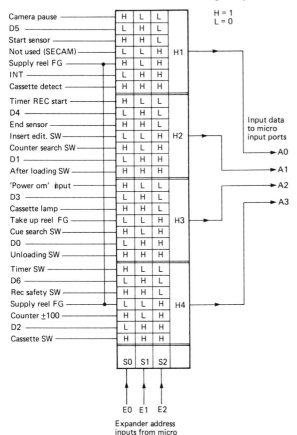

Figure 6.5. Input expander

118

Counter Search Switch, ensures that the recorder stops at 0000 or the tape counter in FF or Rew.

After Loading Switch, closes when threading is completed.

Power on, power up reset to prevent random selection of a function when the recorder is switched on.

Cassette lamp, shuts down all functions if the cassette compartment lamp fails.

Take Up Reel FG, low when take up reel is rotating. If this goes high during play or any other mode then all functions stop, to prevent tape damage.

Cue Search Switch, allows the recorder to stop in FF or Rew if a cue pulse is picked up by the Cue Head. Cue Pulses are recorded each time the machine enters record mode.

Unloading Switch, closed throughout the un-threading process, it is used to rotate the supply spool in the reverse direction to wind the tape back into the cassette.

Timer Switch, puts the recorder into Timer Recording Mode.

Record Safety Switch, prevents any record mode, including edits, if the cassette record prevention tab is removed.

Counter ±100, used in the Counter Search mode, this port is 'high' for a tape counter reading lower than 9900 and higher than 0100, it also goes 'high' at 0000. The effect is to slow down FF and Rew speed within the boundaries of ±100 either side of 0000 to ensure that the tape stops accurately at 0000 without overshooting due to high tape speed.

Cassette Switch, this closes when the cassette compartment has lowered the cassette within the machine. (Note that it is a front loading machine.)

This is not all of the information that is required by the microprocessor, the Drum Servo flip-flop signal and Capstan FG signals are fed directly to two other ports for the protection of the tape. If either of these two signals are absent then all other functions are inhibited.

Function selector

The HR7600/50 System Control microprocessor also has facilities for a wired remote control. The operation of the wired remote control input is similar to that of preceding models HR7200 and HR7300, and is shown in Figure 6.6.

A pulse count from the micro output data ports, G0, G1, G2 and G3, is buffered and mixed. As the four ports give a binary count, a resistive network of 10 k and 20 k resistors is able to provide a staircase voltage scale by adding the four-bit binary pulses. A four-bit binary word is capable of providing 2^4 or 16 steps when added in a resistive adder network. The staircase steps represent voltage levels in sixteen values from 0 V to 10 V and

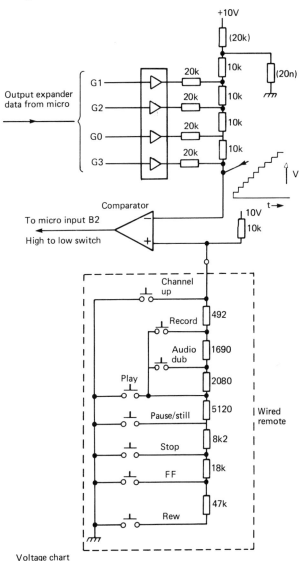

Figure 6.6.

Voltage chart

Channel up	0V
Record and play	0.47V
Audio dub and play	1.7V
Play	2.99V
Pause/still	4.84V
Stop	6.37V
F.forward	7.74V
Rewind	8.92V

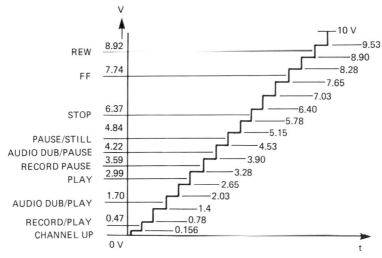

Figure 6.7. *Wired remote control comparator inputs*

these values are inputted to the negative input of an op-amp, used as a comparator. The positive input to the comparator is determined by fixed resistors connected to ground via various function switches. The input voltage is determined by the value of the switched resistors to ground and a 10 k 'pull-up' resistor to 10 V.

With no wired remote control functions selected the 10 k pull-up resistor holds the output of the comparator to a high level. If, say for example, pause/still were selected in play mode, the voltage to the positive input to the comparator would be 4.8 V. The output of the comparator would remain high for the first seven steps of the staircase, that is up to the 4.53 V step. When the binary count on G0 to G3 steps on, the voltage level will change to 5.15 V. This will then make the negative input to the op-amp comparator higher than the positive input, so the output will change from high to low. The change on the input port B2 of the microprocessor will signal a wired remote control selection. The micro will then look at the binary word on G0–G3 and recognise it from its programming as a pause/still command. Still Frame mode will then be entered (see Still Frame and Slow Motion). Output ports G0–G3 also double up as an output data to the output expander, in this respect the output binary from G0–G3 will *not* be a straight count.

In practice a nice neat staircase signal will not be presented to the comparator, and inspection of the waveform will reveal seemingly random voltage levels. This does not matter too much as any binary word on G0–G3 will at any time produce a

voltage reference to the comparator; should any instantaneous reference voltage change the output of the comparator, then the binary word for that voltage will be recognised as a wired remote control command, the micro will then confirm the status by producing a ramp count.

Microprocessor outputs

Direct output controls are available from a custom micro for the Brake Solenoid, Pinch Roller Solenoid, Threading Motor, Cassette Motor and Reel Motor (refer to Figure 6.3).

These are termed direct outputs as they are coded for more than one operation and do not output through the output expander. Solenoid Drives have two parts, one to provide a high current 'operate' pulse and the second to provide a holding current via higher power drive buffer transistors. The threading motor (or Loading Motor) is driven from three ports, F0, F1 and F2 and a bridge network of transistors. This provides for Threading, Unthreading and Motor Braking.

The Cassette Motor, the one which transports the cassette compartment up and down, is also driven by a bridge network of transistors to enable the motor to reverse drive. Four ports are used for this function, D0 and D1 to load, D2 to unload and D3 for motor braking. Two output ports are utilised for the Reel Motor, which drives the spool carriers. This motor also drives forward and reverse, and ports C0 and C1 control the motor direction. Other reel motor functions come from the output expander.

Output expander

The output expander (Figure 6.8) works in the reverse manner to the input expander, as it is similar to a four pole seven way switch again. The output ports G0–G3 are each switched to one of the seven outputs in its own quadrant. The microprocessor address ports S0, S1 and S2 select one of the seven outputs of each of the four sections in parallel.

For example, if S0–S2 were 100 (HLL) then G0 would output to Reel Idler Control, G1 outputs to Reel Motor FF or Rew, G2 outputs to ±100 (FF or Rew) and G3 outputs to Reel motor unloading. All four of these are Reel Motor Control Signals, but the actual output does not appear until the fourth address signal, when a strobe pulse has been inputted to the expander to hold the signal output until the next change is required.

If the Reel Motor is to be in the FF or Rew mode then G1 will be high when S0–S2 is 100, the Reel Motor FF output will be locked high by the next strobe pulse for the duration of the function. A subsequent strobe pulse for the next function used will reset the reel motor FF output to low again.

As S0–S2 and G0–G3 are continuously changing it is necessary to use the strobe pulses to set and hold the output from the expander for the duration of the selected function. It is possible to set more than one output high, as it may be necessary to perform more than one output function at any time, for example, whilst holding the

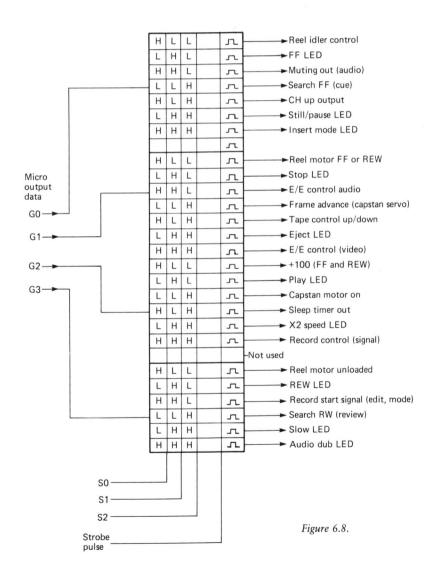

Figure 6.8.

121

reel motor in FF, the FF LED will have to remain illuminated also.

Let us have a look at some of the outputs from the expander:

Reel Idler Control. The Reel Motor Idler pulley normally sits in contact with the Reel Motor Shaft but not with either spool carrier turntable. An idler function is a short high level pulse to spin the Reel Motor and hence the idler pulley in order to 'throw' the pulley over to one of the spool carriers so that it may then be driven by the motor.

LED outputs are self explanatory, to illuminate an LED.

Muting, Audio muting in FF, Rew, Cue and Review.

Search FF (Cue). Search Rew (Review), Cue and Review signals to capstan servo microprocessor.

CH UP, advances the Tuner channel selector in up count 0.5 Hz/s rate.

Reel Motor FF or Rew, runs reel motor at high speed, approximately 45× Normal Play, FF or Rewind is selected by micro ports C0 or C1.

E/E Control Audio and Video, switches Video output and VHF output between E/E (Stop or Record) and Playback signals.

Frame Advance, puts Capstan Servo Control micro into Frame Advance.

Tape Counter Up/Down, digital tape counter counts Up in all forward modes and Down in all reverse modes.

±100 (FF and Rew), in FF or Rewind this signal appears when the tape counter is ±100 either side of 0000, it slows the reel motor down.

Sleep Timer Out, instructs Tuner Timer micro which times down from 60 minutes or a presettable time and then stops tape. Used when 'Timer' is selected whilst the recorder is in Record Mode.

Record Control, switches Audio and Video circuits between Play and Record.

Reel Motor Unloading, used during unthreading to drive supply spool in reverse to take up tape slack.

Record Start, used to inhibit Recording until the correct instant required, used in assembly and insert edit modes.

This is only an overall view of a sophisticated systems control. The circuit details are much more complex and of course the manufacturer's service manual will have to be referred to for these details. It has been used as an example to illustrate the types of switching signals involved and how these signals are matrixed in and out of the control microprocessor.

7

SLOW MOTION AND STILL FRAME TECHNIQUES

The main problem encountered with video recorders when a still frame is to be displayed is that the video heads do not follow the same path across the tape when it is stationary as was recorded by them when it was moving. The magnetic tracks are recorded on the moving tape, and provided the tape is moving during replay the video heads will read a path parallel to the one recorded. If the tape is then held stationary the replay path is *not* parallel to the recorded path and the video head will wander off track somewhere in the replay field. Add to this the fact that each video head will only reproduce its own recorded track, due to the azimuth slant between the gaps and a problem occurs. When the tape is stationary *both* video heads follow the *same* path, so if the path is over Ch.1 track then Ch.2 head will only replay noise. The change in video head paths between stationary and moving tape is shown in Figure 7.1.

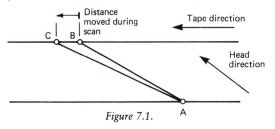

Figure 7.1.

Consider that the tape is stationary and both video heads will follow the path AC: the angle of this track is the same as the angle of tilt of the ruler edge around the drum assembly upon which the tape travels. The distance BC is the amount that the tape travels, when moving, during the time it takes a video head to travel across the tape, following the path AC, during one field scan. When the tape is stationary the video heads contact the tape at point A and travel up, across the tape to point C. If the tape is moving, the video head still contacts the tape at point A and travels towards C, however, during this time, the tape

moves the distance BC. When the video head has completed the scan, C is no longer at the end of the track, B has moved into its place. The path of the video head when the tape is moving is AB.

We therefore have two possible video head paths across the tape. AB when the tape is moving, which is the normal record/replay path. AB is the standard track angle and length for each TV field during normal operation. AC is the path which *both* video heads will follow across the tape when it is stationary. When a standard recorded tape is halted during replay then a path of AC is followed and it will cross over at least one and, at worst, possibly two, of the recorded tracks AB. It is not possible for a video head to replay its track without crossing over to an adjacent track and producing a band of noise. The noise is referred to as tracking error, or more correctly 'crossover noise'.

Crossover noise is not the only distortion which occurs in the replay of a still picture. It can be seen from Figure 7.1 that the track AC is longer than AB; the drawing is an exaggeration for clarity, and in fact the increase in track length is 0.48%, or about three TV lines. If you think about this point carefully you can rationalise that the video head will travel this slightly extended track in the same time period. In other words AC is scanned in the same time period as AB. So 315½ lines are reproduced in 20 ms, indicating a small increase in the replayed line frequency rate, which is so in fact.

For a video head to replay a longer track within the same time period means that the head to tape speed has increased by 0.48%. All replayed frequencies off the tape are therefore subject to a 0.48% increase. Such a small increase in line sync frequency and replayed colour undercarrier can be accommodated by the AFC loop within the colour circuits, but it is inconvenient for slow motion replay, as will be seen later.

Why standard heads cannot reproduce a still picture

Figure 7.2 is taken from the JVC HR3660 service manual. The shaded tracks are Ch.1 tracks and the light tracks Ch.2 tracks. In ①, the tape has stopped in a position where two video heads cover a Ch.1 track. This allows Ch.1 video head to reproduce its track to almost full level, as the FM output shows. Ch.2 video head, following the same path, reproduces only a small amount of FM at the beginning of the scan and at the end with nothing in between, except noise.

In ②, the replay path covers both Ch.1 and Ch.2 recorded tracks; this is still insufficient to enable a noise free replay. Ch.1 video head picks up no FM at the start of scan and almost full level at the end whereas Ch.2 replays full level at the beginning of scan and little or no FM at the end.

path will cover any recorded tracks at random, but also that there is no position that can give a noise free replay. This can be readily seen as true in such models as the Sanyo VTC9300 (Beta) and the later JVC VHS recorders HR7200 and HT7300 which have 'picture pause' rather than still frame.

The VHS solution

A solution to the problem is to modify the video heads. This can be done in three ways: (1) Increase the thickness of the video head chips so that they can cover a wider area of tape in still picture replay. (2) Modify the action of static video head chips so that they can move in the vertical plane and follow the original recorded tracks. (3) Add extra heads.

The first option is the one adopted by JVC, who have put considerable design effort into both still

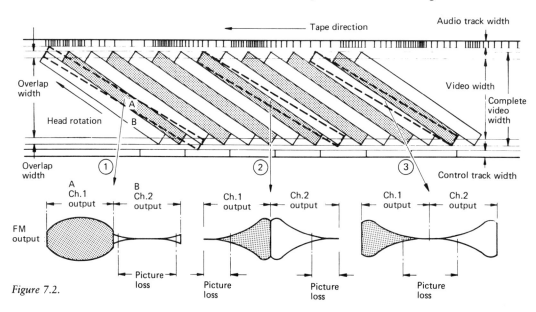

Figure 7.2.

Ch.1 reproduces noise at the top of the screen and Ch.2 reproduces noise at the bottom of the screen.

In ③, again the replay path covers both Ch.1 and Ch.2 recorded tracks. Ch.1 provides an output at the start of scan and nothing at the end. Ch.2 has no output at the start of scan and full output at the end; a situation which will reproduce noise bars at the top and bottom of the picture as in ②.

It proves therefore that stopping the tape during a replay cannot provide for a noise free still frame, not only due to the fact that the replay video head

frame and slow motion. The result of this effort first emerged in the HR3660 video recorder, in which the video heads are modified. The width or thickness of Ch.1 head is increased from 49 μm to approximately 59 μm and Ch.2 head is increased to approximately 79 μm, as can be seen in Figure 7.3b.

An important point to note is that the lower edges of the video heads are on a reference plane. This reference is the same height above the main tape deck plane as the 49 μm video heads in a standard recorder. In still frame operation the

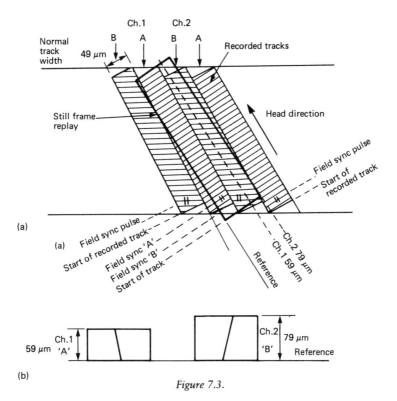

Ch.1   Ch.2
B   A   B   A

Normal
track
width

49 μm

Recorded tracks

Still frame
replay

Head direction

Field sync pulse
Start of
recorded track

(a)

(a)   Field sync pulse
Start of recorded track
Field sync 'A'
Field sync 'B'
Start of track

Ch.2 79 μm
Ch.1 59 μm

Reference

Ch.1
59 μm   'A'

Ch.2   79 μm
'B'   Reference

(b)

Figure 7.3.

servo tracking control system halts the tape under strict control such that the replay track path sits over Ch.1 (shown as 'A') recorded track. Alignment of the replay still frame track path is adjusted by a slow motion tracking control. The user adjusts the slow motion tracking control for a noise free picture; in doing so the replay track path is sat upon Ch.1 track.

The reason for this alignment is illustrated in Figure 7.3a. The replay path is shown in dark outline upon the Ch.1 and Ch.2 recorded tracks shown shaded. Each video head follows the same path, aligned on the reference line, Ch.1 head has a replay track 59 μm wide and Ch.2 head has a replay track 79 μm wide. The result is that Ch.1 head replays recorded track 'A', it is well covered and sufficient FM is picked up to satisfy the minimum input requirements of the replay limiter amplifiers.

Ch.2 video head follows along the same reference line; its extra thickness provides for a very wide replay track, the whole of the area within the dark border. It allows Ch.2 video head to replay the recorded track B and again sufficient FM is replayed to allow a noise free replay. However the limits of the system design are reached at the end of the scan. Generally this will not necessarily

allow still frame replay of a tape made on another recorder.

Techniques using extra thick video heads are used on most makes of VHS recorders which have still frame or slow motion facilities. An exception is the Panasonic NV366, which has three video heads.

Synthesised vertical sync pulse

When the video recorder is in normal play each video head, in turn, replays the FM signal which contains the field sync pulse. The drum servo, in record, has ensured that each field pulse has been consistently recorded at the same point from the start of the track, shown by the ″ symbol, in Figure 7.3a.

In replay, the field sync pulse is replayed with consistent timing after the start of each track. However, in still frame operation both video heads follow the same path and as such they commence the start of scan at the same point, the start of the track in Figure 7.3a.

Ch.1 replays field sync 'A' and Ch.2 replays field sync 'B'; it can easily be seen that the sync pulse replayed by Ch.1 is later than that of Ch.2. The difference in timing of the replayed field sync

125

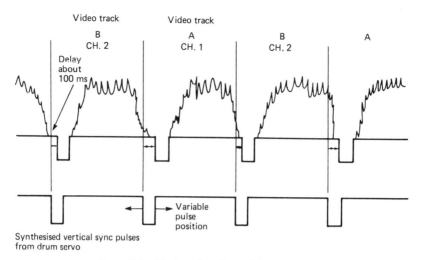

Figure 7.4.   Timing of synthesised field sync pulses

pulses, if allowed, will cause the replayed picture to jitter in the vertical plane, i.e. frame bounce. In order to compensate for this a synthesised field sync pulse is injected into the video signal during *all* trick replays – Still Frame, Slow Motion, X2 and Cue and Review.

The timing of the injected field sync pulse is 100 μs before the video sync pulse in Ch.2 replay. Figure 7.4 illustrates the timing of the replayed pulses and synthesised vertical sync pulses. The timing of the Ch.2 synthesised sync pulses is 100 μs before the replay field syncs. TV sets will lock to the first synthesised pulse and ignore the secondary replayed one.

Synthesised vertical sync pulses are derived from the flip-flop signal which is in turn derived from the rotating video head drum. The synthesised sync pulses will therefore provide for a stable replay in still frame operation. To ensure a bounce free picture the vertical sync pulse for Ch.1 replay is variable in timing, from an external pre-set control.

Recording with different thickness video heads

Once it has been accepted that the basis of VHS still frame techniques is increased thickness video heads, the question that will be asked is 'how can these video heads record normal 49 μm wide magnetic tracks?' The secret is in the fact that the bottom edge of each of the video heads is on the same reference plane.

In Figure 7.5 the first (1) magnetic track is laid down by Ch.A video head, this track is 59 μm wide, initially. Ch.B video head subsequently

scans the tape to lay track Ch.B1 79 μm wide. However when Ch.B video head lays this track its bottom edge is the reference line which is 49 μm from Ch.A reference. The extra 10 μm of Ch.A track is over-recorded by the subsequent Ch.B video head. Again Ch.A video head scans for the second time laying Ch.A2 track. The bottom edge

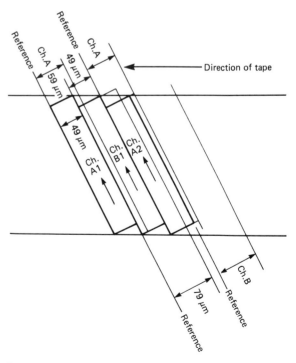

Figure 7.5.   Recording magnetic tracks with extra thick video heads

126

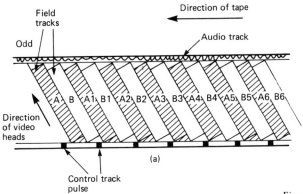

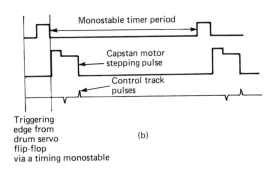

Figure 7.6.

of Ch.A video head is still on the reference plane and the reference line is 49 μm from Ch.B reference line. It lays a 59 μm wide track, and in doing so, over-records the excess 30 μm of Ch.B1 track.

The sequence then repeats throughout the recording, each video head sits on the reference plane and on each subsequent scan the reference edges are 49 μm apart. Each video head thus over-records the excess recorded by the previous one; the magnetic tracks are recorded 49 μm wide as does normal head assembly with 49 μm wide head chips.

Slow motion

JVC use a principle in their video recorders which is not found in any other VHS manufacture in the same format. It is based within a BA841 large scale integrated circuit.

When still frame is selected the BA841, under control of the slow motion tracking control, ensures that the replay track path sits upon Ch.1 video head track as previously shown.

Slow motion is a series of still frames in sequence under the control of the BA841 IC. The rate at which still frames are sequenced is controlled by a slow motion speed control, part of a monostable time constant. At the end of a monostable timer period, when the monostable is reset, the drum flip-flop is used to form a stepping pulse for the capstan motor.

Figure 7.6b shows the basic pulse timing although in practice a few more monostable delays are used, including slow motion tracking. The small pulse is the short monostable reset period, short because once the monostable is reset, the next available flip-flop pulse will set it again. In doing so, the flip-flop is used to develop a castellated stepping pulse.

A DC motor requires a higher starting torque than it does a maintaining voltage, hence the castellated shape of the drive voltage to the motor. Once the drive pulse is initiated it is maintained until a control track pulse causes it to reset. The tape then rests for the remaining monostable timing period until the sequence is repeated and the tape is stepped again, to the next control track pulse.

Slow motion is therefore a series of still frames, and each still frame is displayed for the monostable timer period. At the end of the timer period the capstan motor steps to the next control track pulse. A sequence is shown in Figure 7.6a where the first frame consists of odd field A and even field B.

After stepping the video heads will read off odd field A1 and even field B1. The sequence will be repeated and the capstan motor will step the tape from control track pulse to control track pulse. The slow motion replay will be formed frame by frame and will follow the field sequence AB, A1B1, A2B2, A3B3, A4B4, A5B5 etc. These are true still frames consisting of an odd and an even field and maintaining an interlaced still picture.

The slow motion tracking control affects the start and stop edges of the stepping pulse and in consequence ensures correct 'framing' of the replay video head tracks upon the recorded video tracks. Noise-free slow motion is therefore obtained.

The slow motion tracking control also ensures a noise free still frame. When still frame or play pause is selected during play, the tape does not stop immediately. The BA841 stops the tape transport in four little hops, these hops helping the slow motion tracking control to correctly place the replay tracks over the recorded tracks so that the still frame picture is noise free.

127

Speed play

In Speed play (×2) mode of operation the 59 μm and 79 μm thick video heads also play a major part in providing for a noise-free picture. As the tape is now running faster than it was in record, the replay video head path is at a greater angle than the recorded track. Figure 7.7 shows the speed play situation. Each video head now passes over a pair of recorded tracks because the tape is running past the video head assembly at twice the recorded speed.

by National Panasonic and Hitachi is called visual search.

The visual search system is used to move the tracking noise bar, present in the picture when still frame is selected, out of the picture to leave a noise-free still frame. Figure 7.8 shows the visual search circuit of the Hitachi VT8000 series, with the kind permission of Hitachi.

Reference is also made to the waveform timing diagrams in Figures 7.9 and 7.10. One of the functions of the visual search circuit is to derive synthesised vertical sync pulses for the display TV,

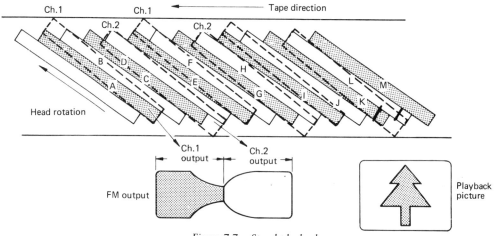

Figure 7.7. Speed playback

Ch.1(A) video head passes over recorded tracks A/B. The extra thickness of Ch.1 video ensures that it replays A field track and reproduces sufficient FM for a noise-free picture.

Ch.2(B) video head subsequently scans the tape and passes over recorded tracks C/D. The extra thickness of Ch.2 video head ensures that it replays D field tracks and reproduces sufficient FM for a noise-free picture.

The recorded tracks A, B, C, D, E, F, G, H etc. are replayed at ×2 speed in the sequence A, D, E, H, I, and L in order to maintain interlace. Over a large number of recorded fields passing through in ×2 speed only half of them are replayed in a 1, 4, 5, 8, 9, 12 sequence for odd and even field interlace purposes.

Visual search technique

Other partners within the VHS system do not have access to the BA841 IC for slow motion and still frame noise-free replays. Other methods have therefore been devised. One of these methods used

as replayed vertical syncs are subject to timing errors in trick modes. One of the synthesised vertical syncs must be in advance of replayed vertical syncs and alternate ones must also be advanced and be externally variable in position to minimise picture bounce. Timing for the pulses developed in the visual search circuit, comes in the form of the flip-flop (or 25 Hz switching signal) on pin 8 of PG1101.

Synthesised vertical syncs

Derivation of the variable vertical sync pulses is accomplished by a NOR gate in IC1101 pins 11, 12 and 13. The incoming flip-flop signal is buffered and inverted in IC1104/12, and the output is connected to IC1101/12. Further inversion of the flip-flop is done in another stage of IC1104/4 and fed to IC1101/13 via an integrator circuit comprising R1101 and C1103; this will delay the rise time of the square wave to IC1101/13. Comparison of IC1101/12 and 13 are shown in the timing diagram.

128

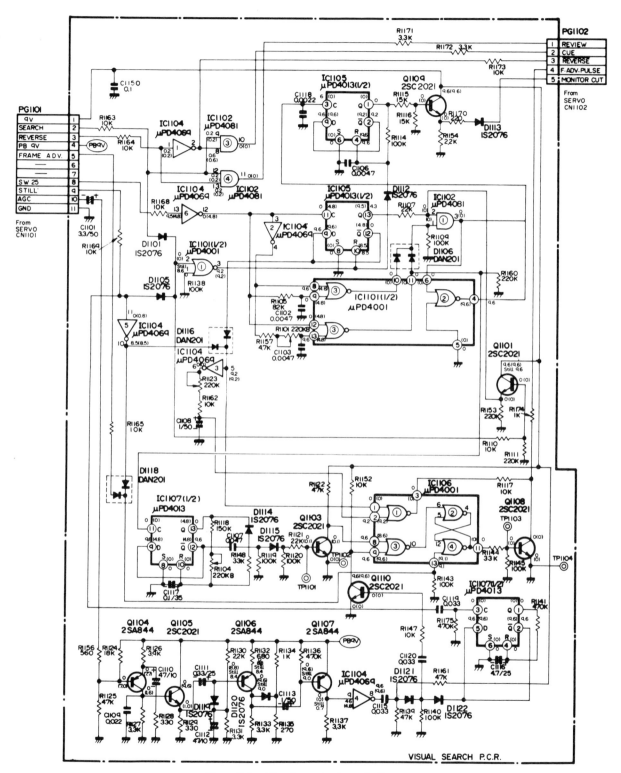

Figure 7.8. Visual search schematic diagram

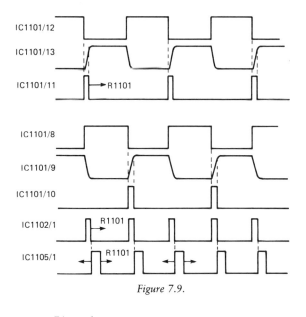

Figure 7.9.

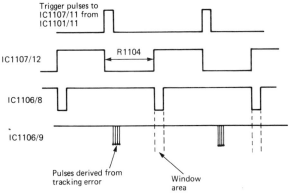

Pulses derived from
tracking error

Window
area

Figure 7.10.

The output of IC1101/11 is high when both inputs are low, and as pin 13 rises above the IC threshold the output changes to low. The width of the output pulse is therefore dependent on the time constant C1103 and R1101 where R1101 determines the position of the trailing edge. A second section of IC1101, pins 8, 9 and 10, produces the fixed vertical sync pulse timing by the same procedure: inversion of the input signals is required to determine the pulse position alternate to the variable pulse position. Both outputs from IC1101 are added by diodes to IC1102, a NAND gate, where pin 2 is held high by Q1101 in all trick modes.

IC1105 (pin 3) is triggered by the trailing edges of the combined pulse outputs from IC1102. IC1105 is a monostable which has a fixed timing period, which is the synthesised vertical sync pulse period. The output of IC1105/1 is shown in the timing diagram Figure 7.9. The variable trailing edge position determined by R1101 will therefore vary the synthesised vertical sync pulse position out of IC1105/1. The derived pulses from IC1105/1 are buffered out by Q1109 on the 'monitor cut' output pin 5, PG102.

The term 'monitor cut' is used to refer to the video output muting circuits, synthesised vertical sync pulses are inserted into the video signal by pulse driving the muting circuits.

Visual search operation

The aim of the visual search circuitry is to determine a pulse 'window' at the end of each video head scan, that is, at the bottom of the replayed picture; also to derive a suitable pulse signal from the tracking noise which is on the screen, and then shuttle the tape such that the two pulses coincide. Trigger pulses for IC1107/11 are the ones already derived as drive pulses for the synthesised vertical sync pulses, the leading edge is used to trigger IC1107.

IC1107 is a monostable with a period variable around 20 ms by R1104, and the positive edge on pin 12 is the monostable reset edge. A pulse is derived from this reset edge by the time constant C1107/R1119 turning on Q1103, the output being connected to the NOR gate IC1106/8. The derived pulse is the 'Window' for a gating system and the position of the window pulse is variable. Note that the pulse is at frame rate and therefore the visual search operates over two field tracks to locate Ch.1 video head onto Ch.1 video track in still mode. An input to the visual search circuit on pin 10 of PG1101 is from the AGC and dropout compensator.

The circuit comprising of Q1104–Q1107 amplifies and limits the tracking noise to produce a bunch of pulses via Q1110 to IC1106/9. Whilst the pulses to IC1106 pins 8 and 9 are *not* coincident Q1108 is off, frame advance pulses are derived in the NOR gate pins 1 and 2 of IC1106; the width of each of the pulses is variable by R1123. The output pin 3 of IC1106 feeds the drive pulses out to the capstan servo, as frame advance pulses.

Once still frame has been selected frame advance pulses are produced and the tape forward shuttles until the two inputs to IC1106 pins 8 and 9 are coincident, Q1108 is then turned 'on' and the frame advance pulses are inhibited.

By adjusting R1104 the position of the tracking bar can be adjusted to a point where it is not visible in the picture. Frame advance is achieved by turning Q1108 off by pin 13 of IC1106; the tape then shuttles until the input pulses coincide once again.

Visual search is used in the Betamax video recorders by Toshiba in the V5470 model. This recorder has a very similar system to the one just described.

Sony C7 also has a visual search system. Again the basic system is similar in so much as a window pulse is generated to be in a position at the bottom of the screen. The noise bar, which in the C7 is shuttled to the bottom of the screen, is identified by a colour phase error signal developed in the colour circuits rather than utilising the replay FM carrier noise.

Toshiba V8600

A Betamax solution to noise-free still frame and slow motion first appeared in the Toshiba V8600, the first domestic video recorder to have *four* video heads. Figure 7.11a illustrates the position of the four video heads on the drum assembly.

Ch.1 and Ch.2 video heads are standard Betamax record/replay heads for normal replay. Ch.2A and Ch.2B are non-standard video heads, thicker than the normal video heads, similar to the VHS slow motion technique. In still frame mode Ch.1 and Ch.2 video heads are switched off and the special Ch.2A and B video heads are switched on. The tape transport is shuttled so that the Ch.2 heads are situated over a Ch.2 recorded magnetic track, as shown in Figure 7.11b.

The extra thickness of the video heads gives a replay path which is much wider than the recorded track. This provides for sufficient coverage of the recorded track to replay the FM signal without any losses due to the change in recorded and replay track angle.

In slow motion the capstan servo is pulsed to step the tape from control track pulse to control track pulse with a slow motion tracking control to ensure that the special Ch.2A and B heads are optimised on the Ch.2 tracks. In ×2 speed, again the extra thickness of the Ch.2A and B heads is utilised to provide a noise free replay. In point of fact, the Toshiba still frame is in reality a Still field

as the same magnetic track is repeated, by only a Ch.2 replay. Ch.1 magnetic track is not used in any of the still frame or slow motion modes.

A function of the V8600, not widely known, is its 'record pause' editing capabilities. If pause is selected during the recording mode the tape transport does not stop until Ch.2 head has finished scanning the tape. Also when the systems control

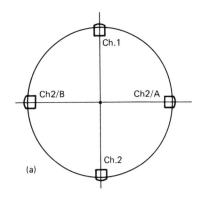

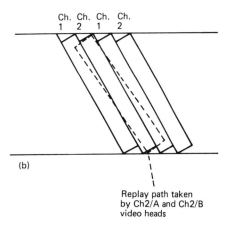

Figure 7.11. Toshiba V8600 four-head system

is taken out of record pause the tape transport does not recommence until Ch.1 video head is starting to scan. With the electronic/mechanical timing involved within the V8600 the result is a very clean assembly edit; there may be a small amount of chroma disturbance but this is normal in VHS also.

V2000 still frame

A solution to the problem of change in the angle of video head to tape between normal play and still frame mode can be found in the V2000 type of video recorder which does not require a change in video head thickness. In fact, the technology available to this system should have produced a more simplified approach. The video heads are mounted upon piezoelectric crystals and can be moved in the vertical plane by the application of a high voltage.

Automatic tracking was one achievement from the use of dynamic tracking frequencies but corrections derived from these alone would not be enough to compensate for the change in track angle between moving and stationary tape. Figure 7.12a shows the recorded magnetic path compared to the uncorrected replay path in still frame mode.

In order to 'bend' the crystal mounting, a ramp voltage is applied to it; the ramp is produced by the DTF microprocessor program. It can be considered as coarse correction, the DTF system providing fine correction on top of the ramp, as shown in Figure 7.12b. A positive voltage bends the video heads in the direction of the tape whereas a negative voltage bends the video heads against the direction of the tape. Note that, if uncorrected, both video heads, Ch.1 and Ch.2 will follow the same path in still mode. At the beginning of Ch.2 head scan maximum correction voltage is required; this reduces, as Ch.2 scans, down to 0.

Ch.1 head starts along the same path and initially requires little correction voltage. As Ch.1 scans, the negative correction voltage is gradually increased to a maximum at the end of Ch.1 scan. This enables the recorder to satisfactorily replay a still frame with interlaced odd and even fields. The relationship between the ramp correction voltage and the amount of shift required for Ch.2 and Ch.1 heads, and the direction of the shift is shown in Figure 7.12c.

Ch.2 video head, at the start of its scan path, need to be shifted towards the left of the diagram, in the direction of the tape travel. The initial high positive voltage applied to the piezo crystal will cause this shift. As the head continues its scan the need for correction reduces and so does the positive correction voltage.

At the end of Ch.2 scan, the video head is 'on track', the correction voltage is zero volts. Ch.1 video head starts its scan on track, so the correction voltage remains zero. As Ch.1 video head

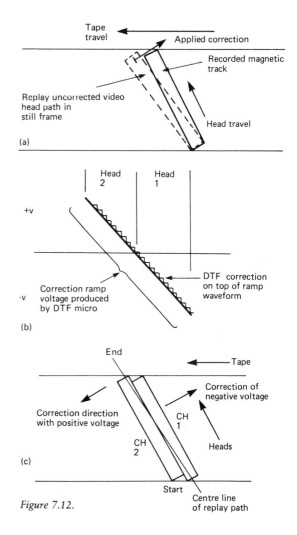

Figure 7.12.

scans, following the centre line of the replay path, the need for correction to the right arises, this is a negative voltage. As the Ch.1 video head completes its scan the negative and correction voltage will be at its greatest value to shift the head towards the right, and on track.

Still frame selection

Figure 7.13 illustrates the selection of an appropriate pair of magnetic tracks for a still frame noise-free picture. There are four possible selections of Ch.1 and Ch.2 magnetic tracks: f1/f2, f2/f4, f4/f3, and f3/f1. The point at which the still frame is selected is not the point at which the tape stops, due to inertia. Even when the tape transport stops, it may not be an ideal pair of tracks, due to higher than normal correction voltages being applied to the crystals.

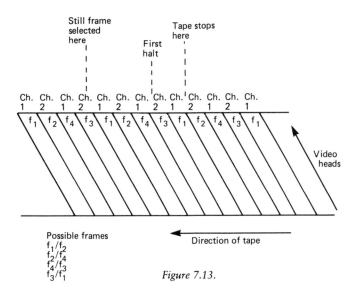

Still frame selected here

First halt

Tape stops here

Ch. Ch. Ch. Ch. Ch. Ch. Ch. Ch. Ch. Ch. Ch. Ch. Ch.
1   2   1   2   1   2   1   2   1   2   1   2   1

f_1 f_2 f_4 f_3 f_1 f_2 f_4 f_3 f_1 f_2 f_4 f_3 f_1

Video heads

Possible frames
f_1/f_2
f_2/f_4
f_4/f_3
f_3/f_1

Direction of tape

Figure 7.13.

The microprocessor control system monitors the correction voltages; if they are too high then further tape transport is initiated. The first halt is on Ch.1 and Ch.2 tracks with DTF frequencies f4/f3; if the correction voltage is too high the microprocessor issues DTF frequencies f3/f1, which will also progress the tape transport. The microprocessor will then continue to step the transport further by changing the DTF outputs until a suitable pair are found with only minimal DTF correction upon the preset ramp voltage, in this case f1/f2.

It is possible, therefore, for the microprocessor based DTF system to not only select a pair of magnetic tracks in still frame, but also to optimise for minimum DTF control voltages.

Slow motion

It is now fairly clear that the microprocessor based DTF system can dictate which magnetic track a video head scans by issuing the DTF frequency for that track. This principle is utilised in the slow motion mode. The tape speed is reduced to one third normal play, as shown in Figure 7.14a. The microprocessor then issues DTF frequency f1 for Ch.1 head, f2 for Ch.2 head and f4 for Ch.1 head again in sequence. The sequence f1, f2 and f4 is then followed by repeating the latter two, that is f2, f4 and f3 and so on.

If we pick up from f3 which is a Ch.2 video head, the next track to be replayed is two back, that is f4, Ch.1. From Figure 7.14b, we can see

that for Ch.1 replay a large positive control voltage is applied to displace the video head to the left to put it on Ch.1, f4 in the third sequence. This is followed by Ch.2, f3 and the correction voltage falls to zero. A negative ramp control voltage will then put Ch.1 video head forward to track Ch.1, f1.

Repetition of this sequencing maintains Ch.1 and Ch.2 odd and even field interlace. By playing three fields and then repeating two, the replay only advances by one field each time, this being one third speed replay.

Cueing at seven times forward speed

The ability of the V2000 DTF system to maintain noise-free track following can be extended to its limits of 7× forward cueing with no tracking errors. This is achieved by the microprocessor dictating every seventh track whilst the tape transport runs at seven times normal play speed in the forward direction. The sequence is Ch.1/f1, Ch.2/f3, Ch.1/f4, Ch.2/f2, Ch.1 and Ch.2 interlace is maintained on every seventh track, as shown in Figure 7.15a.

Figure 7.15b illustrates the difference between the actual track and the replay path. At the start of scan a negative voltage is required to shift the video head to the right and at the end of scan a positive correction voltage is required to shift the video head to the left. This is supported by Figure 7.15c, which shows a positive going ramp voltage.

Tape direction

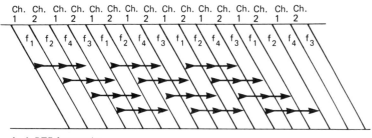

Ch. Ch. Ch. Ch. Ch. Ch. Ch. Ch. Ch. Ch. Ch. Ch. Ch. Ch. Ch. Ch.
1   2   1   2   1   2   1   2   1   2   1   2   1   2   1   2

f_1 f_2 f_4 f_3 f_1 f_2 f_4 f_3 f_1 f_2 f_4 f_3 f_1 f_2 f_4 f_3

f_1–f_4, DTF frequencies

(a)

Figure 7.14.

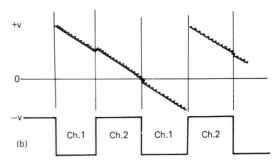

+v

0

−v

Ch.1   Ch.2   Ch.1   Ch.2

(b)

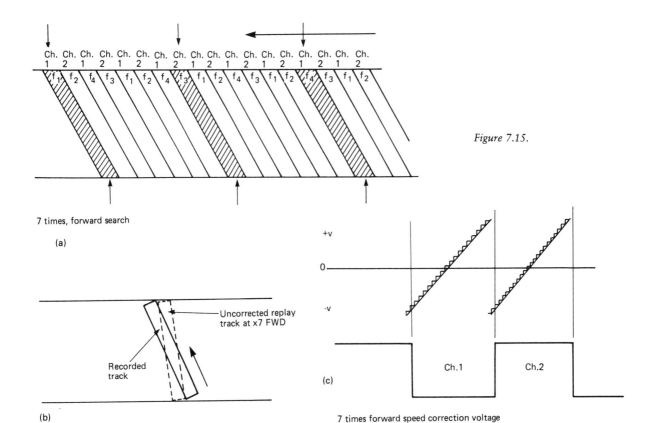

Ch. Ch. Ch. Ch. Ch. Ch. Ch. Ch. Ch. Ch. Ch. Ch. Ch. Ch. Ch. Ch. Ch. Ch.
1   2   1   2   1   2   1   2   1   2   1   2   1   2   1   2   1   2

f_1 f_2 f_4 f_3 f_1 f_2 f_4 f_3 f_1 f_2 f_4 f_3 f_1 f_2 f_4 f_3 f_1 f_2

7 times, forward search

(a)

Figure 7.15.

Uncorrected replay
track at x7 FWD

Recorded
track

(b)

+v

0

-v

Ch.1   Ch.2

(c)

7 times forward speed correction voltage

134

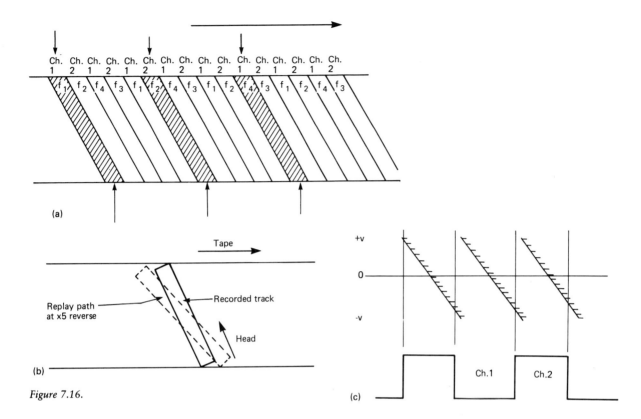

Figure 7.16.

Review at five times reverse speed

With the tape transport running at 5× reverse speed the microprocessor selects every fifth track in the sequence as shown in Figure 7.16a. The sequence is Ch.1/f4, Ch.2/f2, Ch.1/f1 and Ch.2/f2 which is then subsequently repeated.

Figure 7.16b illustrates the difference between recorded and replay track paths. At the beginning of scan a shift to the left is required which is a positive correction voltage. At the end of scan a shift to the right is required and the correction voltage is negative.

Figure 7.16c shows a negative-going ramp which passes through zero with minor corrections upon it.

JVC VHS microprocessor controlled slow motion and still frame automatic tracking

This is a later development from JVC for selection of a noise-free still picture. As previously shown in Figure 7.6 a capstan stepping pulse is used to ensure a noise-free still frame picture. Adjustment

for such a noise-free picture is achieved by a variable slow motion tracking control.

The limitation of a slow motion and still frame tracking control is that it requires readjustment for each different cassette. This adjustment has to be carried out in the slow motion mode, even for still frame. A later system from JVC is found in the HR7650, where the capstan servo is tied to a microprocessor and the old BA841 LSI chip has been discontinued.

In Figure 7.17 the operation of the system is shown by the timing pulses. In ① after the pause (still frame) is selected the microprocessor counting delay period (a) is initiated by the first positive edge of the flip-flop waveform. This corresponds to Ch.1 video head commencing replay of its track, after the video head crossover point. After the delay count period (a) a motor reverse pulse (b) is used to stop the capstan motor dead in its tracks by applying a reverse pulse voltage. There then follows a further flip-flop counting period whilst the microprocessor checks the FM signal, if all is well a standard motor drive castellation pulse is used to step the capstan motor.

A 'standard' drive pulse has a start edge, period (c), after a Ch.1 positive flip-flop transient. This is

135

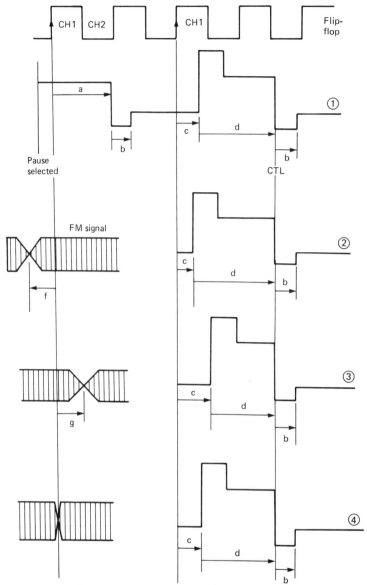

Figure 7.17. Automatic slow motion tracking

followed by a standard run period (d) which is terminated by a braking pulse (b). However, if during a subsequent checking of the FM signal, the crossover point is prior to the Ch.1 positive transient, then the tape is judged not to have travelled far enough. This is shown in Figure 7.17 ②.

The crossover point is converted to a pulse from the FM AGC circuit, and a measurement is taken (f) between the crossover point and Ch.1 video head start of scan. The period (f) is then subtracted from the (c) period and the start of the motor drive pulse is advanced. Run period (d) is therefore

longer and tape travel is increased to make up for the shortfall.

In section ③ of Figure 7.17 the crossover point is found to be after the Ch.1 reference edge by period (g). In this instance the tape has travelled too far and a smaller motor drive step is required to correct for it. The microprocessor therefore adds the period (g) to period (c) in order to lengthen it. Run period (d) is therefore reduced and the motor drive pulse is shortened.

In section ④ the crossover point coincides with the Ch.1 video head transient, which means that

136

the tracking noise is off the screen at the bottom of the picture. A noise-free picture is therefore obtained and the stepping pulse is then re-adjusted to standard.

When a still frame is selected, the microprocessor has a maximum of eight stepping pulses to set up a noise-free picture. If it fails to do so it will stop on the eighth anyway.

In slow motion mode, continuous checking and correcting is carried out to maintain a noise-free picture.

The system is automatic for still frame and slow motion tracking and it will operate for interchanged tapes as well as the replay of the video recorder's own tapes.

Editing

For some years much development has taken place in the direction of electronic editing. Editing on U-matic recorders has reached a point where it is carried out between two video recorders and a microprocessor based control unit. Whilst editing on VHS equipment has not yet reached such a level, it is available on certain models.

Editing is not carried out with video tape in the same manner as film. Cutting and splicing are most definitely out. Even the most carefully spliced video tape will have a joint which, to the small video head, will look like a large ridge. The video head will suffer mechanical shock against the joint and in many cases will suffer damage. Editing is therefore done electronically by a method which ensures a noise-free joint between previously recorded and new material. Two methods are employed which we shall look at individually; these are Assembly Editing and Insert Editing.

Assembly editing

Assembly editing is joining new information or a new recording onto the end of a previous recording. This can be done in two ways. One is to record the first section and then maintain the video recorder in 'record pause' whilst sorting out the second section, then allowing the recorder to continue. Let us call this the 'record pause' method. The other way is somewhat more complex, in that the first section has already been recorded and it is reviewed on the video recorder in the Play mode. The point at which the new information is to be added is selected and the

recorder is put into 'play pause' and then into 'record pause' and then straight record.

The success of either of these two methods depends on the degree of technology involved in the video recorder being used. The most successful assembly edits (and for that matter inserts also) are achieved on video recorders that employ 'flying erase heads'. Using Figure 7.18 as a model you can see a small amount of tape between the erase head and Ch.1 video head. It is this section of recorded tape that creates the problems.

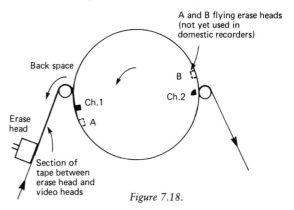

Figure 7.18.

If we assume that the model is a U-matic editing video recorder, it will have flying erase heads, shown as A and B. These precede the Ch.1 and Ch.2 video heads. In any editing mode these flying erase heads will switch on and erase all video information on the tape just before Ch.1 and Ch.2 video heads record new magnetic video tracks. The advantage is that the full erase head is not switched on and it will leave the audio and control tracks alone.

A video recorder fitted with flying erase heads will be able to 'video dub' over recorded audio information, changing the video signal but leaving the audio track existing. Unfortunately domestic video recorders do not yet have this facility.

Back space editing

Back space editing is the standard technique used on most domestic video recorders. When record pause is selected, the video recorder rewinds the tape for a very small portion before entering 'pause' mode. All of the section between the erase head and the video heads has already been erased during the normal recording process. After the pause, the recorder will continue in record mode to overlap the previous recording with the new recording for only about half a second or so.

Some video recorders do not backspace but rely on instantaneous stopping and starting of the tape by timing the pinch wheel action. In the small overlap period between the recordings the electronic 'joint' is quick and clean, that is to say there is no picture break up or loss of servo stability; there is, for reasons to be detailed later, a small amount of colour flicker.

A worse colour flicker situation occurs when the 'play pause' is selected and the whole section of tape between the erase head and the video head has recorded information upon it. The video recorder will backspace by a small amount before continuing in record. In this instance the overlap period is much longer, some 3–4 seconds before tape cleaned by the erase head reaches the video heads.

During the overlap recording period, the video heads will over-record the old information; this is effective for the luminance FM carrier but not so effective on the colour undercarrier which is apparent in the background for the duration of the overlap. This overlap period, caused by recorded information being present on the tape in the section between the erase head and the video heads, lasts as long as it takes the beginning of the newly erased tape to travel up to the video heads.

Insert edit

In the new era of domestic video recorders, Insert edit is the capability to insert new audio and video signals into previously recorded information.

This has to be achieved with a clean start to the new recording without interference and picture break up and with a clean ending, back to the original recording. An example of the uses of Insert editing would be to insert a title into previously recorded programme material.

JVC HR7650 assembly edit

The assembly edit system of the JVC HR7650 is identical to the previous HR7700 model except that timing values are slightly longer. In the HR7700 the backspacing is 20 frames with play-

back of 12, which gave the HR7700 12×40 ms = 480 ms to prepare for the edit point, which in most instances is only just enough time to lock the capstan servo to incoming video.

Figure 7.19 illustrates the HR7650 assembly edit system, and its function sequence. Initially the video recorder is in the normal record mode. Pause is selected, identified by the systems control microprocessor as 'Record pause'. It therefore enters the 'Edit' mode as follows. Upon selection of 'Pause' the tape stops. Under the control of the microprocessor the tape rewinds 25 frames which is 1 second of programme material, achieved by counting 25 control track pulses in rewind. The tape is then held at stop in 'Record pause'. The Slow/Still overtime counter is initiated for a 5–7 minute count, after which the tape is unthreaded to prevent any physical damage. Editing operations therefore have to be completed within this time period.

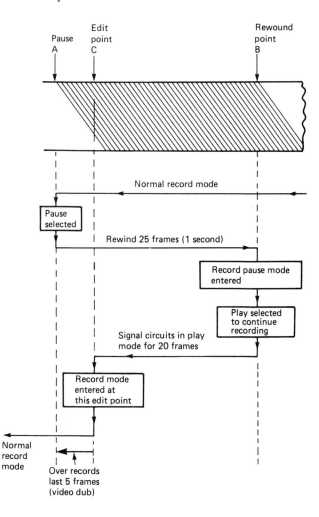

Figure 7.19.   Assembly edit

When the information to be assembled onto the tape is ready, further operation of the video recorder is done by selecting the 'Play' button. A period of 20 frames is then entered with parts of the machine in Play and parts in Record. The E/E signal is monitored at the output of the recorder, which is a record function, signal circuits are in record, *BUT* record is inhibited and the video and audio heads are not recording. 20 frames is an interim period in which the capstan servo is in 'Play' mode and it uses replayed control track pulses as feedback. Incoming vertical sync pulses, suitably divided by two are used as the reference input to the capstan servo. The reason for this is similar to jointing sections of film together. As the film is cemented and overlapped the sprocket holes are aligned on top of each other.

During the 20 frame period the capstan servo phases up replayed control pulses to incoming vertical syncs that will be responsible for recording the new control track pulses after the edit point. At the edit point, which is selected from the ending of the 20 frame period and a subsequent first Ch.1 flip-flop transient, the capstan servo will switch from Play to Record. The idea being that, during the 20 frame period when incoming vertical syncs are phased up to replayed control track pulses, the capstan servo ensures that the first new control track pulse recorded is exactly 40 ms after the last one replayed prior to the edit point (see Figure 7.20). When the edit point is subsequently replayed there is no timing error between the control track pulses over the edit point, only the video and audio signals change.

After the Edit point, there are 5 frames of information remaining, before blank tape erased by the erase head. These 5 frames are erased by the video heads over-recording them. FM carrier is erased or re-recorded but an amount of colour under carrier will remain. This can sometimes be seen as a colour flicker, lasting only 200 ms. The audio track is erased by an audio erase head, next to the audio head, and is not subject to audio crosstalk between new and previous recordings when the edit point is subsequently replayed.

Insert edit

Insert edit was not available on the HR7700 but it is on the HR7650, the method being shown in Figure 7.21. First, the Edit out point is selected, this is the end of the insert, so we are working backwards. It is selected by zeroing the digital tape counter to 0000. The Edit In point is then found by using picture search rewind, reversing the tape with picture monitoring until a suitable point is found. Play or slow motion can also be used, the point is then selected by pressing the 'Pause' button and obtaining a still frame replay. Insert Edit mode is then entered by pressing both 'Pause' and 'Insert edit' which is recognised by the microprocessor as an edit sequence. As in the Assembly edit mode the tape is rewound 25 frames where it sits in Record pause.

Upon selecting 'Play' the recorder advances in the Pseudo-play mode for 20 frames allowing the capstan servo to phase up incoming syncs to replayed control track pulses. After 20 frames the edit point is reached and insert recording takes place. Unlike assembly edit, only video and audio recording takes place, the capstan servo remains in Play with the drum servo in Record; this enables the capstan servo to maintain operation using control track pulses as feedback and incoming syncs as reference. During Insert edit *NO* new control track pulses are recorded; provision is

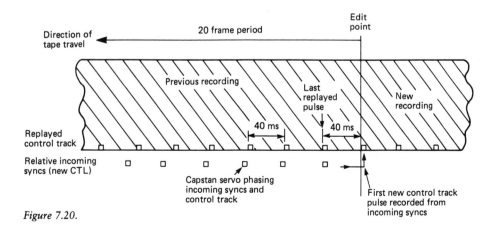

Figure 7.20.

139

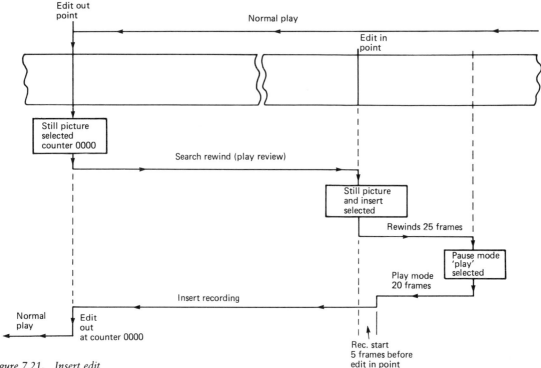

Figure 7.21. Insert edit

made within the system to record pulses if replay-ed control track pulses go absent.

When the counter reaches 0000, the Edit out point, Insert edit is completed and the recorder reverts to Play; this is timed by the RS flip-flop from the drum servo to prevent visual interruption by switching after the head crossover point. The full erase head is not used during an Insert edit. Audio is erased by an audio erase head prior to the new recording.

Luminance FM carrier is over-recorded by the video heads, the FM record signal is boosted by 15–20% during the edit period to use the video heads as erase heads by increasing the recording flux. The S/N on replay is reduced by this method as the recording current is no longer optimum. Chrominance crosstalk will also be noticeable under certain conditions and it will be seen as a coloured flicker in the background.

It is best to insert edit high contrast pictures onto low contrast material to avoid recognisable crosstalk. If dark scenes are inserted into bright scenes, then the insert edit may not be considered to be of good quality.

Servo signals

The inputs to drum and capstan servos on a video recorder with additional facilities, such as still, slow motion and editing will not be as normal. A resumé of the servo operation is as follows.

Figure 7.22 shows the two servos drum and capstan with the signal inputs for play and record. Operation in search and slow motion is not shown for clarity.

Inputs to the drum servo in Record mode are vertical syncs, divided by two to 25 Hz (40 ms period), which is the reference, and the PG pulses for feedback from the motor. There is, in addition, the FG feedback loop for speed control.

The reference input forms the ramp and feedback is the sample for the phase lock side of the servo. This is standard for a drum servo in record. In playback, feedback is still PG but the reference is a crystal oscillator, suitably counted down. An important point for purposes of editing is that the reference input to the capstan servo is the ramp signal from the drum servo. Feedback in the capstan servo is the FG signal divided down to 25 Hz in record and control track signal in replay.

In Record, both servos reference from vertical syncs, the drum directly and the capstan servo indirectly via the drum servo. They are, then, locked to incoming syncs, the drum servo provides control track recording drive.

In Playback, the drum is referenced to a crystal oscillator, so is the capstan. The drum servo runs at constant speed and phase, whereas the capstan servo has feedback from the replayed control track, it also embodies the tracking control. A variation occurs during editing. In the 20 frame

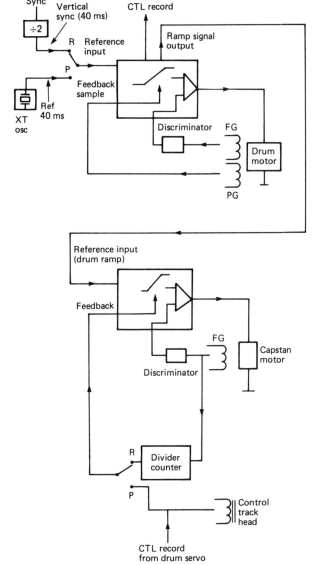

Figure 7.22.

replay period the drum servo remains in Record, the capstan servo switches to Play. This means that the ramp reference in the capstan servo is derived from the drum servo and is from incoming syncs. Feedback at this time is from replayed control track pulses, hence the capstan servo is able to phase lock incoming syncs and replayed control track and has 800 ms to achieve phase lock. The capstan servo in Assembly edit reverts back to record after 20 frames, whereas during Insert it remains in the play mode. At the end of an Insert edit period the drum servo is switched from record mode back to play mode.

It is necessary for an Assembly/Insert edit video recorder to have control over signal record by inhibiting during editing and to be able to switch drum and capstan servos individually between Record and Play.

This complexity of control systems is available at low cost in domestic machines due to microprocessor programming techniques and mass production.

As the recorders develop for the domestic market we can expect to see even more complex production versions, without tuners and able to be controlled by editing suite microprocessor control units.

Still field techniques

In this chapter we have looked at the reproduction of still pictures by scanning A & B tracks generally with the use of a modified B channel head to cover the B track. Two problems are created by this technique. First, the B channel head is inefficient and the signal to 'crosstalk' noise is very low, in standard play the signal to noise ratio is reduced also, thereby degrading the picture. Second, the reproduced picture is a still frame i.e., two consecutive fields are repeated and so any fast movement that has occurred within the time of the two fields is displayed as a jitter.

The solution is to display a single *field* and repeat it then, as there is no movement, there can be no jitter. There *is* a reduction in vertical resolution, however, as only 312 lines are displayed, but it is considered that a reduction in vertical resolution is a better trade off than jitter.

Various methods have been used to achieve display of a still field, including the use of two or four video heads, but the final solution is a digital field store.

Three video heads

The three head solution to still field was widely used by Panasonic in VHS and Sony in Betamax. The Panasonic three headed assembly is shown in Figures 7.23 and 7.24. In Figure 7.23 the two main record and replay heads A & B are shown 180° apart, the third head is designated B' and has the same azimuth tilt as B. It cannot be mounted in the same place as the A head for practical reasons and so it is mounted in front of the A head by 620 μm. In the still field mode the two heads B & B' follow the same track: they are both 70 μm wide to compensate for the change in track angle between play and stop as illustrated earlier in this chapter. However, the B' head is ahead of its counterpart as they are not 180° apart and a shift in horizontal syncs will occur between each field and cause severe horizontal displacement. To compensate for this loss of horizontal correlation the B' head is lower by 15 μm and by triangulation puts it in the correct timing position to restore correlation. This ensures that horizontal timing is stable but vertical timing will have a two line shift giving rise to vertical jitter. To get round this, in all modes other than record and play, synthesised vertical sync pulses are employed and timed from the rotating head drum. By introducing a timing delay into one of the synthesised field pulses compensation is achieved, and by making the delay variable with a 'vertical lock' control, the user can adjust for minimum jitter. A circuit is also included which detects the presence of any noise bars in the picture period and steps the capstan motor to remove the noise bar to the field blanking period and, so, off the screen.

Practical construction of the video head cylinder is shown in Figure 7.24 where the B' head is mounted in the same assembly as the A head. Azimuths for the heads and their track widths are shown also. The signal to noise ratio is slightly degraded by the use of 70 μm heads in a system with 49 μm track widths but this is improved by careful design of the FM amplifiers.

Four video heads

In a higher grade videorecorder, the NV366, Panasonic improved overall performance by introducing a drum cylinder with four video heads in a similar manner to the Toshiba V8600. Construction is shown in Figure 7.25. There are two main heads designated A & B with track widths of

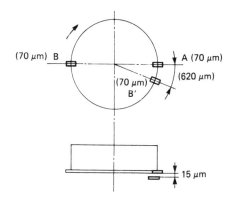

Figure 7.23.

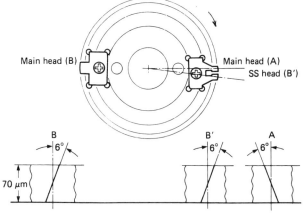

Figure 7.24.

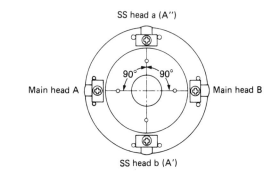

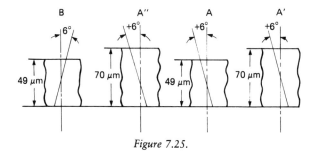

Figure 7.25.

49 μm to obtain the best signal to noise ratio for normal recording and playback. There are also two heads, mounted 90° to the main heads, designated A' & A''. These two additional video heads have the same azimuth of +6° but are 70 μm wide. They will provide for a high performance still field with no loss of horizontal correlation and minimum vertical jitter. Synthesised vertical sync pulses are used to prevent loss of vertical sync due to the noise bars being shuttled into field blanking.

Combined four heads

In most general types of videorecorders in the picture search or cue and review modes of working, the pictures are broken up by noise bands. The noise occurs when the video heads pass over a video track with an opposite azimuth to the heads, and happens to both video heads. Each video head will cross over 9 or 10 tracks during a single field scan in picture search when the tape speed is 9–10 times normal play. JVC has produced some very high performance video recorders and went to some lengths to improve the still picture and slow motion performance. In doing so the picture search was also improved and the noise bands reduced to just thin lines which did not greatly detract from the picture information. This was done by the use of four video heads, two for normal record and playback, referred to as the SP (standard play) and two for LP (long play). As the LP heads were used to 'support' the two main video heads the noise-free picture search applied only to standard speed.

Figure 7.26 illustrates the head mounting configuration. SP Ch.1 head and SP Ch.2 head are the standard record and replay heads and are spaced 180° apart, SP Ch.1 head has track width of 59 μm and SP Ch.2 is about 70 μm wide. The LP Ch.1 and LP Ch.2 heads are also 180° apart but displaced from the main heads by a distance corresponding to 2 TV lines. While SP Ch.1 and SP Ch.2 are used for standard speed normal recording and playback, in picture search or still picture mode all four heads are active. It is possible to do this as the azimuths of the LP heads are opposite to the SP heads with which they are co-mounted so that SP Ch.1 and LP Ch.2 are together, and SP Ch.2 and LP Ch.1 together. This

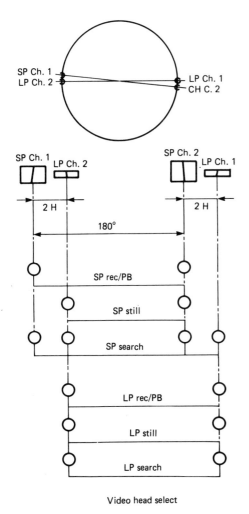

Video head select

Figure 7.26.

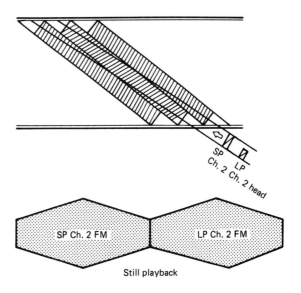

Still playback

Figure 7.27.

143

puts SP Ch.2 and LP Ch.2 180° apart except for the 2 line displacement. The reproduced picture is a still field (312.5 lines) from SP Ch.2 and LP Ch.2 heads reading the same video track as shown in Figure 7.27. Picture vertical jitter created by the 2 line displacement is compensated for by a 2 line shift in the vertical synthesised sync pulse timing periods between the drum flip flop switching signal and the synthesised vertical sync pulse.

A more interesting feature of the use of four video heads is in the SP picture search mode of operation when all four of the video heads are brought into use. By switching between the four video heads selectively it is possible to replay sufficient FM carrier on a continuous basis to eliminate noise bands, as shown in Figure 7.28. In sector 1, the Ch.1 video head signal starts off low and increases, whereas the Ch.2 starts off high in level and decreases. By selecting a level about half of maximum and then switching from Ch.2 to Ch.1 the output signal level can be maintained. In this example Ch.1 and Ch.2 signal can be from either the SP or the LP heads. Figure 7.29 shows

the block schematic of a microprocessor-controlled switching system for noise-free picture search.

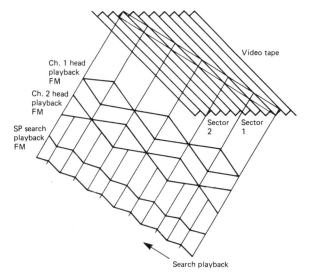

Figure 7.28.

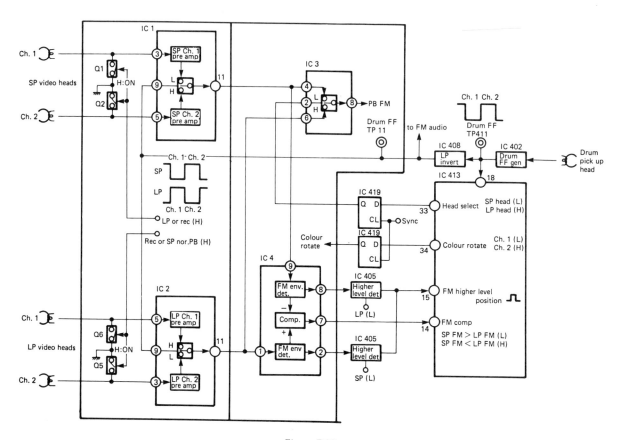

Figure 7.29.

ICI switches the outputs of the SP video heads and IC2 switches the LP heads. When the drum flip-flop output is low in level, SP Ch.1 and LP Ch.2 are 'active'. When the flip flop output is high then SP Ch.2 and LP Ch.1 are selected. The selected outputs from ICI and IC2 are passed on to IC3. Switching between the SP and LP heads is done in IC3 by a 'head select' signal which is high for LP and low for SP and so only the output of ICI for SP heads, or IC2 for LP heads, is output as playback FM carrier. With a combination of head select and flip flop switching pulses any one of the four video head outputs can be routed through the playback system. The transistors Q1 and Q2 are used in record to provide an earth path for the video heads and preamp inputs, the record FM signal being applied to the other end of the video head winding. Transistors Q5 & Q6 perform the same function in LP record mode.

The head select switching signal is synchronised to line syncs by IC419 because switching the video heads at the beginning of a TV line minimises picture disturbance. This is of little use in normal playback but is significant in the noise-free picture search performance. Also in a part of IC419 is a colour rotate switching signal, derived from the head switching signals to ensure the correct colour phase, again more important at the head switching points during picture search and is derived from the head select signal within IC413.

Flip-flop switching signal is generated from the drum PG pick up head as discussed in Chapter 4. IC402 and IC408 invert the flip-flop signal between SP and LP. The inverted signal in LP mode is also delayed by a 2 line period (128 μs) to compensate for the physical displacement between the two sets of heads upon the mounting assemblies.

When picture search is selected IC4 becomes active, comparing the FM envelope levels of the SP and LP head outputs. The output of the comparator is low if the SP heads' FM signal is greater and high if lower than the LP heads.

Microprocessor IC413, controls the switching functions and is programmed for the noise-free picture search. In slow motion or still picture modes it controls the stepping of the capstan motor by determining the noise bar position within picture area. This is done by FM measurement from IC4 and comparing the event time within the confines of the drum flip-flop period before stepping the noise bar to the vicinity of a flip-flop transition i.e., out of picture area.

In picture search mode the output levels of all four heads are effectively checked and the highest output switched through with switching synchronised to line syncs and colour phase rotated through 180° to provide for the minimum of picture disturbance at the switching point.

Digital field store still picture

Signals obtained from video heads are analog in nature. To store an analog video signal in a digital memory, it must first be sampled and converted into a binary code. The theory of analog-to-digital (A/D) conversion is fairly straightforward; in Figure 7.30a we have an example of a small portion of analog video signal and in Figure 7.30b this signal is sampled eleven times. Eleven samples is not very many; the sample video signal has curves but only 11 points *on* the curve have been measured or sampled. If we were to try to reconstruct that sample of video signal in a digital-to-analog (D/A) converter we would get the result shown in Figure 7.30c out of the sample and hold circuit, which does not really resemble the original signal due to the low number of samples. Clearly, the resolution of the A/D converter is not

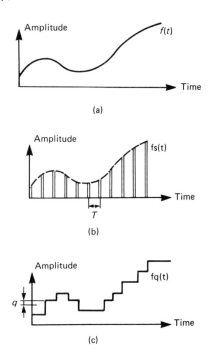

Figure 7.30. a, original signal. b, sampled signal. c, quantised signal

145

good enough and so to digitise and then reconstruct the sample video waveform accurately many more samples would have to be taken. As the number of sample points is increased the reconstructed waveform would begin to look more like the original with smaller and smaller steps in the waveshape.

Obviously, there are limits as to the number of samples that can be taken and, as a guide, the sampling rate has to be more than twice the highest frequency component of the signal. For a colour video signal the normal minimum sampling rate is taken to be three times the chrominance subcarrier frequency:

$$= 3 \times f\text{sc}$$
$$= 3 \times 4.43618\,\text{MHz}$$
$$= 13.3\,\text{MHz (approximately)}.$$

Next comes the question of the vertical size of the steps. Distortion of the signal due to the introduction of small steps is called 'quantisation error'. In order to minimise quantisation distortion and maintain resolution of a video signal it is sampled and digitised in as many values of steps as possible. Generally, 64 values are sufficient.

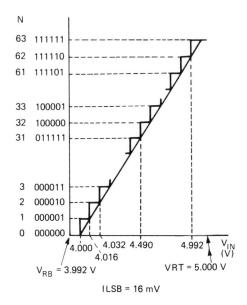

Figure 7.31.

up of 64 voltage steps. A filter is used to remove the steps and restore the signal to a more rounded form.

6-bit A/D conversion

The video signal is a standard 1 V p/p waveform so the minimum level is 0 V and the maximum level is 1 V, but for technical reasons, when converting from analog-to-digital video signal is clamped to a defined DC level, say, 3.992 V for sync tip. For a 1 V p/p signal peak white will be 4.992 V. This means that the input video signal will vary between 3.992 V and 4.992 V.

A 6-bit binary word from 000000 to 111111 has 64 different counts so it is possible to have 64 grey scale levels from 3.9992 V to 4.992 V. The amplitude of the signal will therefore be quantised into 64 levels and each level will have its own 6-bit digital word. Figure 7.31 illustrates the conversion characteristic where 000000 is equivalent to 3.992 V and 000001 is the binary code for 4.000 V. Each increase of 16 mV returns the next binary count so the quantisation of the video signal is in 16 mV steps. Each step is numbered from N=0 to N=63 for the 64 counts of a 6-bit data word.

D/A conversion is the reverse of the above, where each digital word will produce a voltage level on any one of 64 levels to reconstruct the original video signal as a stepped waveform made

Memory integrated circuits

In this example of a digital field store, used within a Toshiba DV80, the memory consists of DRAMs (Dynamic Random Access Memory) and six are used for the field store memory. DRAMs have a high integration which leads to high speed, however they do need constant refreshing to maintain the memory cells fully charged as there is a tendency to leakage. Each DRAM has a memory configuration of 64K by 4 bits, a total of 256 K memory bits.

In order to provide address data for 64,000 locations a 16-bit data address is required but an 8-bit microprocessor cannot directly produce this. It is common practice to use 8-bit data address lines twice, 8 bits to determine a row address and then 8 bits for a column address as shown in Figure 7.32a. In this way the memory locations can be considered to be in a square array of 256 × 256 locations and an 8-bit address can determine which out of 256 rows and then a further 8-bit address determines the column. The memory cell at the intersection of the row and column is the chosen cell.

The memory ICs used for the field store have four data terminals and there are four sets of 64K

146

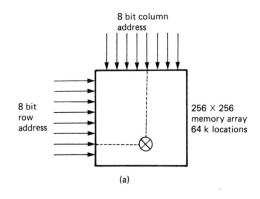

(a)

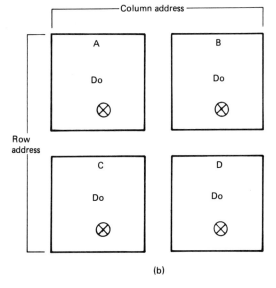

(b)

Figure 7.32.

MB81464–12PSZ
(bottom view)

Pin name	
$A_0 \sim A_7$	Address input 8 bit
$\overline{CAS}$	Column address strobe
$1/0_1 \sim 1/0_4$	Data 1/0 4 bit
$\overline{RAS}$	Row address strobe
$\overline{WRITE}$	Read/write input
$\overline{OE}$	Output enable
VCC	Power supply (+5 V)
VSS	Ground

Figure 7.33.

memory cells whose address lines are wired in parallel. Therefore for any given row and column address the same memory cell location is determined in each of the four 64K memory arrays and one bit of data is stored in each array (Figure 7.32b). The effect is identical to having four 1 bit × 64K memory ICs in a single package.

Figure 7.33 shows the pin connections and terminal allocations for the ICs used. In practice, the row address data must be present at the same time as the RAS (Row Address Strobe) pulse is present, changing to the column address data to correspond with CAS (Column Address Strobe). To determine a memory location the address bits are used twice for RAS and CAS. The significance of the bar above RAS and CAS on the diagrams is to indicate that the voltage level is normally high and goes low for the strobe pulse. The read/write input is low for 'write' and high for 'read'.

Memory capacity

The sampling frequency has been determined by the minimum for video of $3f_{sc}$, the quantisation number is 6 for the 64 levels of samples and the time for the full memory is 20 ms, the field rate. Required memory capacity is calculated from the formula:

M = sampling frequency × time × quantisation number.

The result is that 1,596,103 bits of memory are required for a total field store.

The memory capacity of one of the DRAMs used is 256K, which in precise terms is 262,144 bits (i.e. 2^{16}), and so the total memory capacity of 6 DRAMs is 1,572,864 memory cells. It is then fairly obvious that there is not quite enough memory capacity for a full field store, and only 308 lines of the field can be stored. Figure 7.34 shows the practical solution. The full address capacity will readout 308 lines, and the microprocessor is programmed to jump back to address 32768 and read out four lines up to address 33631. These four lines are added onto the end of the field to make up 312 TV lines. This rarely causes any visual problems as the last four lines are usually out of sight, due to normal TV overscanning.

To prevent colour problems a small circuit is used (Figure 7.35) to remove the colour from these four additional lines by outputting a pulse for the duration of the repeated read to switch in a low

147

pass filter consisting of CV18 and QV25. The whole unit is added to the conventional video recorder chassis as a separate fully screened module.

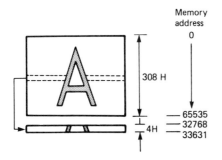

Figure 7.34.

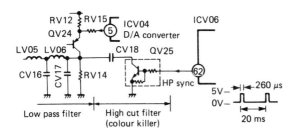

Figure 7.35.

6-bit to 4-bit conversion

When handling video signals timing is always a problem which has to be kept in mind and it is also so when processing digital signals. In the process of A/D conversion of the video signal the sampling rate of $3f_{sc}$ produces very high data rates. The 6-bit data obtained from sampling has a clock rate of 13.3 MHz, which is a cycle time of 75 ns. This means that every 75 ns a 6-bit data word appears on a 6-port output as parallel data. Unfortunately the DRAM memory ICs have a memory writing speed of 240 ns, subject to ±20 ns tolerance.

A method of slowing the data down is used which also falls in line with the 4-bit data ports of the memory ICs, and is shown in Figure 7.36. Data from the A/D converter occurs in words of 6-bits every 75 ns, and four of these words are clocked by the 13.3 MHz clock timing signal into a shift register and latch system so that a total of 24 data bits are latched. A 3.33 MHz clock is derived by dividing the $3f_{sc}$ clock by 4, and it shifts out the 24-bit block of data as a 24-bit parallel word. Obviously, the shift-out rate is one quarter of the shift-in rate i.e., about 300 ns which is well within the writing time of the memory ICs.

Each of the memory ICs holds four data bits out of the 24 available and the 4 × 6-bit data words out of the A/D converter have been converted to 6 × 4-bit data words in the memory array. Each of

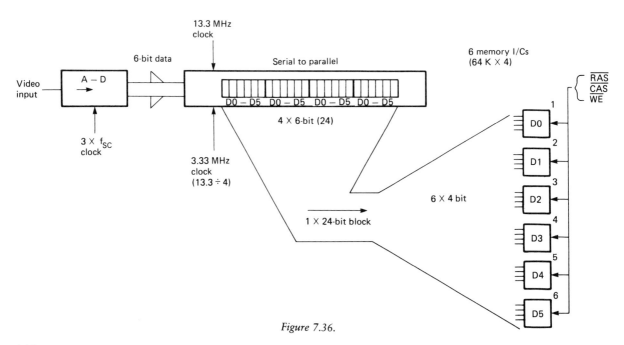

Figure 7.36.

148

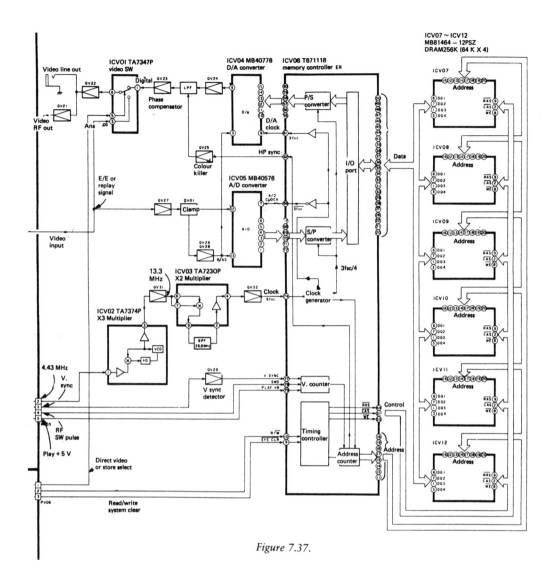

Figure 7.37.

the 6-bit data words is spread out over the six memory ICs, the first IC holds D0, the second IC holds DI etc. As each IC has four separate memories it will store the same data bit for each of the four 6-bit words. IC1 will store four D0 bits, IC2 will store four D1 bits etc.

Following groups of 4 × 6-bit words are converted to 6 × 4-bit words in the same manner, and stored sequentially until the memory is full and IC1 holds 262,144 D0s with the other ICs holding 262,144 D1s, D2s, D3s, D4s and D5s.

It is now fairly clear that an 8-bit microprocessor with a standard 64K address capability can only store 308 lines and the cost to expand to 16-bit processing just for four lines would not be cost effective in a domestic VCR.

Overall system

There are a number of inputs to the digital store. Main signal is input video which can be either replayed video or E/E video: E/E video signal will allow for off-air still pictures as well as from the off-tape replay.

Timing signals necessary for operation are: V syncs, RF flip-flop switching, and play +B. These form the inputs to a vertical counter to determine the start of the field and time up with the flip-flop to 'grab' a field off-tape when still pause mode is selected before the tape stops.

Three signals: read/write; system clear; direct/store select, are derived from the system control logic of the VCR control microprocessor, system

149

clear is used upon power up to clear spurious signals from the store.

Chrominance subcarrier, f_{sc}, at 4.443618 MHz from the colour circuits is multiplied by three in ICV02 with a phase locked loop to 13.3 MHz. This is then further multiplied by two in ICV03 by filtering the second harmonic which gives the basic clock frequency of $6f_{sc}$ (26.6 MHz). All other clock frequencies are then derived within the memory control IC by suitable division.

Input video is converted to 6-bit digital words in the A/D converter ICV05 and clocked at $3f_{sc}$ into the serial/parallel converter within ICV06 where 6-bit data is then converted to 24-bit data. The output port handles the bi-directional transfer of 24-bit data in and out of ICV06 to the memory ICs.

Row address strobe, column address strobe and write enable are sent to control the memory ICs out of pins 54, 51 and 53 to synchronise the video data and address data lines for memory storage or reading.

During reading of the store, 24-bit data is converted back to 6-bit data in the parallel/serial converter and transferred out of ICV06 to a D/A converter, ICV04.

The video signal out of QV24 still contains the small quantisation steps in the signal and these steps manifest themselves as a 13 MHz interference carrier. A low pass filter removes the 13 MHz switching signal. QV23 compensates for the high frequency phase shift and buffers the stored field to the output selector switch.

8

VHS LONG-PLAY

Whilst there is a number of varying tape speeds used in Betamax recorders (notably in the USA) all UK Beta machines to date use the standard (1.87 cm/sec) tape speed. During 1983 JVC introduced a long-play version of the VHS format, and 'clones' appeared under the Ferguson banner. LP models were also produced by Panasonic and Hitachi. In order to double playing time by running the tape at half speed (11.7 mm/sec) various problems had to be overcome. The video track width is reduced by half, from 49 microns to 25 microns; the replay signal-to-noise ratio suffers accordingly, and new noise reduction systems were evolved, as described under 'FM recording and replay advances' in chapter 3. To maintain colour stability in long-play picture-search modes special 'jump' circuits had to be provided; luminance signal corrections became necessary to compensate for the lack of horizontal correlation of the LP video tracks on tape.

The form of the LP video track is determined by the slower tape speed and the extra pair of LP video heads fitted to the head drum. In the first long play machines these auxilliary heads were spaced 70° round the drum from the standard play (SP) heads; current practice is to mount LP and SP heads on common ferrite chips.

In the *standard* VHS format video tracks are laid down side by side with a 1.5 line offset between the start points of successive tracks. This ensures that each line in a given track lies next to lines of equal colour phase above and below, and that start edges (line sync pulses) lie adjacent – see Figure 5.3. It is not possible to achieve the same symmetry within the LP mode due to conflict between tape speed and the fixed mechanical track angle, which is necessarily the same as in standard play. Figure 8.1 shows the difference between SP and LP track configuration. It can be seen in (b) that the 0.75 line offset (half the standard 1.5 line offset) results in the adjacent line patterns being displaced. Colour phase correlation between adjacent tracks is also lost; whereas lines 2 and 316 carry the same PAL phase in SP diagram (a) the same two lines in LP diagram (b) are greatly offset.

Conventional colour crosstalk cancellation techniques can cope with inter-track chroma crosstalk, but the extra measures outlined above are required to satisfactorily deal with the luminance carrier crosstalk conditions encountered in LP mode – they take the form of FM carrier interleaving at half line rate, described towards the end of Chapter 3.

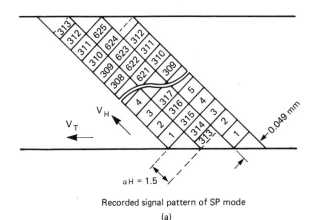

Recorded signal pattern of SP mode

(a)

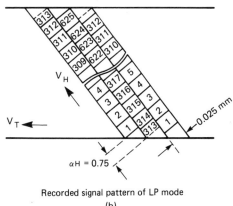

Recorded signal pattern of LP mode

(b)

Figure 8.1.

LP search mode

The main problems occur during picture search when due to the increase in linear tape speed a video head will cross over a number (usually about 5) of its own video tracks as it traverses the tape ribbon. In normal standard play these track crossings create no problem since video track line correlation ensures (given drum speed correction) that continuity of replayed line sync pulses is maintained at 64 μs intervals. In LP search mode, however, each track crossing will give rise to a half-line jump in replayed sync, and if this goes uncorrected the result will be severe picture skew (line pulling) on the monitor TV.

In Figure 8.2 is drawn a section of recorded tape with the upper edge of the ribbon on the right and the lower edge on the left; the slanting tracks are shown as horizontal lines for clarity. The light tracks are those for Ch.1, and the diagonal path is that of the Ch.1 replay head during forward picture search (cue) mode. Shaded lines carry a PAL burst phase of 135° and blank lines 225°.

In the lower half of Figure 8.2 is a timing diagram, row 1 of which shows the off tape replay signal. It is a reconstruction of the replayed lines and colour phases replayed by Ch. 1 head as it reads out the *Ch.1* tracks it encounters. By projecting down from the tape diagram the reason for the disordered signal becomes clear. The replayed TV lines, when demodulated, are not at regular 64 μs intervals; the irregularities are in fact half-line intervals, that is to say the small squares are half a line long.

The line sync timing of the replayed signal is shown in row 3 of the diagram, and this must be corrected to prevent picture skew in LP search modes. It is done by generating a half line (0.5 h) *jump pulse*. A data signal is derived from a phase locked 2 fh oscillator whose output is then divided by 2 to render a symmetrical square wave – row 2 in the timing diagram. It is compared to replayed line syncs in a clocked bistable whose output goes *low* when the data signal is low in comparison to a sync pulse; and goes *high* when the data signal is high in comparison to a sync pulse. This bistable output constitutes the 0.5 h jump pulse (4).

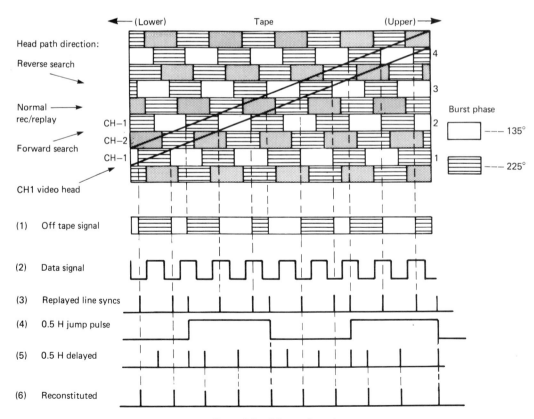

Figure 8.2. LP search mode

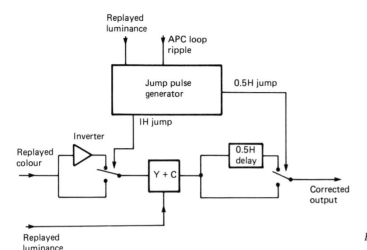

Figure 8.3. *Jump-switching system*

The off-tape video signal is delayed by a period of half a line, and the jump pulse toggles a switch to select direct (3) or delayed (5) video signal as required. The corrected output signal (6) contains no errors to upset line sync on the replay TV.

Two methods have been used to achieve the 0.5 h delay. An early one involved frequency modulating the baseband video signal onto a 14 MHz carrier, passing it through a conventional 32 μs glass delay line, then demodulating it again. Somewhat crude, but entertainment-standard pictures are not required in visual search mode! The second and more up-to-date technique is to use the bucket-brigade device we met in chapter 3. Here a 423 bit CCD serial delay line has the video signal clocked through it at 13.3 MHz rate. A 13.3 MHz clock has a period of 0.075 μs, and 423 transfers at 0.075 μs each comes to 31.8 μs, about half a

line – see Figure 3.34 and associated text for a full explanation.

Obviously if the composite video signal is to be switched at half-line periods, the colour phase must be inverted to prevent the PAL signal becoming disordered and changing phase halfway through a TV line. To prevent this happening and to synchronise the colour phase with the video signal switching, a suitable 1 h jump pulse is generated for the colour signal. It is applied to an inverting switch in the colour replay path.

The complete system is shown in block form in Figure 8.3. Inputs to the jump pulse generator are replayed line syncs (to produce the 0.5 h pulse) and colour APC ripple signal at 7.8 KHz for the 1 h (colour switching) pulse. Usually the whole of the processing takes place inside one or two LSI ICs.

9

HI-FI AUDIO

Audio recording in domestic videorecorders for a long time followed the standard sound recording/replay techniques that are used in any common cassette or reel-to-reel audio recorder. Audio frequency response in conventional videorecorders is limited by the linear tape speed to around 8 or 10 KHz. Stereo videorecorders simply split the 1 mm audio track into two tracks (each 0.35 mm wide) with a 0.3 mm guard band between them. The right hand channel is written along the very top edge of the tape, and any distortion of the tape ribbon edge – caused by dirt, misaligned or worn tape guides or stretching – severely affects reproduction in the right hand channel. Audio level fluctuation, drop out and phase distortion of the right hand channel becomes evident as the wear and tear in a well-used tape takes its toll.

A world market was envisaged for a videorecorder with a substantial improvement in audio performance, brought about by the public's growing appreciation of good sound quality, and by the increasing availability of television receivers with built-in stereo channels using Hi-Fi amplifiers and respectable speaker systems. A high quality audio system in the videorecorder would also make it worthwhile to directly link video machines to the separate domestic Hi-Fi system.

The technical challenge faced by the designers was to improve the audio quality to a level that could be termed Hi-Fi and be comparable to current state-of-the-art Hi-Fi systems and digital compact disc audio players. This had to be achieved whilst retaining system compatibility between the Hi-Fi machine and earlier models, and without affecting video signal quality. A system was developed whereby the audio signals, as FM carriers, could be added to the video signal prior to recording. Although this technique worked with the NTSC system, and found some application in the USA, it was not applicable to the PAL system without severe mutual sound/vision interference.

Research into interference-free audio carrier techniques brought about the method of recording the FM audio tracks by means of separate audio heads mounted on the video head drum. This confers the obvious advantage of high head to tape speed, an increase from 2.339 cm/s to 485 cm/s. The video information is then recorded *over* the audio tracks, partially erasing the audio signal track patterns, but leaving enough for replay. Mutual crosstalk is reduced to an acceptable level by careful selection of the audio carrier frequencies and the use of a large azimuth tilt on the audio heads.

The audio signals are frequency modulated onto RF carriers, one for left and one for right hand channel. The reduction in level of 12 db brought about by over-recording by the video heads does not therefore affect the play back sound quality; the FM signal is unaffected by amplitude variations. A further reduction in crosstalk between the FM carriers of video and audio is achieved by the physical difference of azimuth tilt. With FM carriers replay level drops off rapidly with increasing azimuth offset; whereas the video head gaps are set at ±6° for VHS (7° for Beta) the new

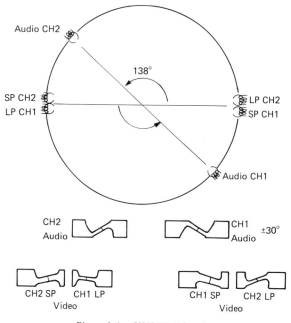

Figure 9.1. VHS Hi-Fi head

154

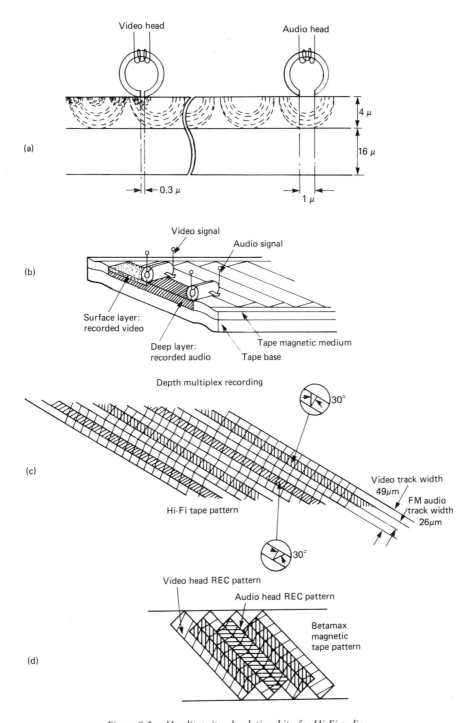

(a)

(b)

Video signal

Audio signal

Surface layer:
recorded video

Deep layer:
recorded audio

Tape magnetic medium

Tape base

Depth multiplex recording

(c)

Hi-Fi tape pattern

Video track width
49μm

FM audio
track width
26μm

(d)

Video head REC pattern

Audio head REC pattern

Betamax
magnetic
tape pattern

Figure 9.2.   Head/tape/track relationships for Hi-Fi audio

155

audio head gaps are cut at angles of $\pm 30°$. Consequently each audio and video head produces an output only when it traces its own recorded track.

Depth multiplex recording

The JVC HRD725 VHS video head drum assembly is shown in Figure 9.1. All video heads are mounted on one line across the drum, with SP Ch.1 and LP Ch.2 together and SP Ch. 2 and LP Ch. 1 together. The reason for the reversal of these heads' position is that in still-picture and slow-motion modes the reproduced picture is a still-field, built up by the alternate replays of SP Ch.2 head and LP Ch.1 head. It can be seen that the audio heads are placed 138° ahead of the video heads and at a height from deck reference level such that the audio head's path lies along the centre of the video track, i.e. the audio and video heads operate in the same plane, see Figure 9.2(c). Whilst the video heads record a 49 μm-wide track, the width of the audio heads and tracks is 26 μm. In standard play, audio Ch. 2 head is allocated to Ch.1 video track, whereas in long play the Ch. 1 audio head is allocated to Ch.1 video track.

Figures 9.2(a) and (b) illustrate the depth-multiplex recording principle. As the audio head has a wider gap than the video head its magnetic field penetrates deeper into the oxide layer on the tape. The surface layer of the audio magnetic pattern is subsequently erased by the following video head on its recording sweep, but the resulting attenuation of the replayed audio signal is only around 12 db, depending on the energy level of the tape in use. Separation of the audio and video carrier signals during replay is dependent on the azimuth difference between the audio and video heads.

Although some of the first Hi-Fi videorecorder models (notably Panasonic types) had stereo capability only on the Hi-Fi tracks and a mono longitudinal audio track, later models from most manufacturers offer stereo in both Hi-Fi and longitudinal modes for complete compatibility with all tapes.

Beta configuration

In Beta Hi-Fi videorecorders the audio FM tracks straddle the joints between adjacent video tracks as shown in Figure 9.2(d). This is necessary because the Beta video heads are 43 μm wide, and write 32.8 μm-wide tracks on the tape – the reasons behind this are explained in Chapter 7. Here a single video head centred on its track will overlap adjacent video tracks on both sides; if audio tracks were aligned along video-track centres, the video head would straddle *three* audio tracks, with consequent risk of visual patterning. The track configuration settled upon ensures that a video head path can only embrace two audio tracks at any time, with reduced risk of patterning. In VHS format this does not seem to be the case; the burial of the audio track under the centre of each video track (Figure 9.2c) works perfectly in practice.

Hi-Fi recording and playback

The left- and right-hand audio signals are frequency modulated onto a 1.4 MHz and a 1.8 MHz carrier respectively, with a deviation of ± 150 KHz. For Beta Hi-Fi the standards are 1.44 MHz and 2.10 MHz with a deviation of ± 500 KHz. In both cases the two carriers are added, then recorded by the audio heads; it is important to appreciate that each audio head caters for *both* sound carriers during its 20 ms sweep of the tape. The Hi-Fi signal spectrum, relative to that of the video signals is shown in Figure 9.3.

A block diagram of the basic record/replay system is given in Figure 9.4. The required programme is routed through the input selector, typically taking the sound channels of a simulcast (simultaneous radio and TV broadcast) from an FM radio tuner, while the videorecorder takes video from its built-in TV tuner. Recording sound level can be chosen to be AGC- or manually-controlled with levels indicated on LED ladder arrays. Muting takes place during 'power up' and 'power down' to eliminate spurious noises before the signal enters the compressor. The function of the compressor is to reduce dynamic range, permitting a wide range of subsequent frequency modification; and considerably improving overall signal to noise ratio. After undergoing reduction of dynamic range and frequency equalisation the audio signal passes through a pre-emphasis circuit and a level limiter on its way to the FM modulator. The limiter prevents any risk of driving the modulator to a deviation greater than 150 KHz. The modulated RF carrier for the left

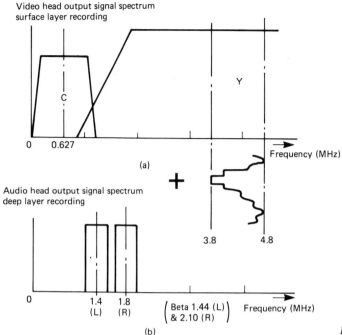

Figure 9.3.   VHS frequency spectrum

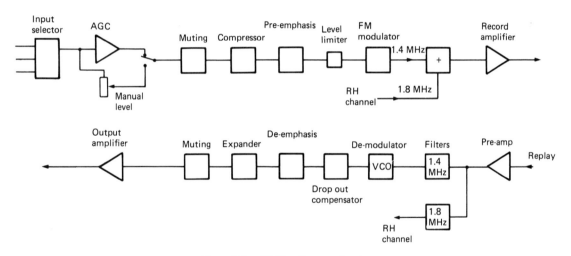

Figure 9.4.   Hi-Fi audio record/replay

(Ch.1) signal at 1.4 MHz is added to the right (Ch.2) carrier at 1.8 MHz before being routed to the audio recording heads via a drive amplifier.

During replay the left- and right-hand signal carriers are individually tuned by bandpass filters for routing to their respective demodulators and drop out compensators. DOCs are necessary to mask interference arising from tape drop out and audio head switching transients. Demodulation is performed by using the recording modulator oscillator as a VXO in a phase locked loop; the 'error' output of the loop forming the recovered audio signal. De-emphasis followed by expansion corrects for the signal modifications carried out during record, reducing noise and restoring dynamic range in so doing. Power muting eliminates spurious noise in the final output signal.

157

Compander

In order to maintain a high dynamic range (and hence signal to noise ratio) of 80 db in the overall record-to-replay signal a *compander* (*comp*ressor-exp*ander*) system is used – Figure 9.5. During record the dynamic level and frequency response ranges are compressed and modified. Correction is applied during replay to restore the signal to its original form in terms of bandwidth and dynamic range. As we shall see, the compression and expansion systems follow a logarithmic law and are both carried out within a specially-developed IC.

In a Dolby system only the high frequency components of the signal are modified; low level high frequency signals are amplified prior to being recorded, and then attenuated during replay to restore balance *and* reduce the level of tape noise.

The compander used in VHS Hi-Fi works on all frequencies and levels, with the frequency response tailored by pre-emphasis networks.

The control range of the logarithmic compressor is shown in Figure 9.6. At the 0 db point it is neither an amplifier nor an attenuator. If the input signal increases by 120 mV the output becomes attenuated by 20 dB, whereas a decrease in input of 120 mV will increase the gain by 20 dB. As is well known the dB scale is a logarithmic one, and the compressor follows its law – for every 60 mV that the input signal falls below 0 dB, 10 dB of gain is applied, and for every 60 mV that the input signal rises above the 0 dB level, 10 dB of attenuation is introduced – compression indeed!

If the compressor worked in this fashion over the whole audio frequency range, the necessary pre-emphasis process that follows may cause over-deviation of the FM modulator stage in the

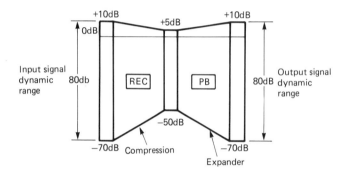

Figure 9.5. *Noise reduction*

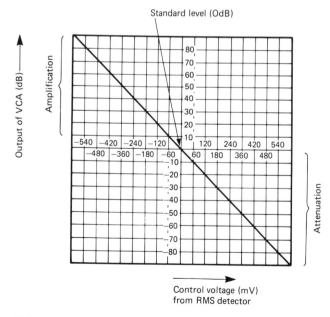

Figure 9.6.

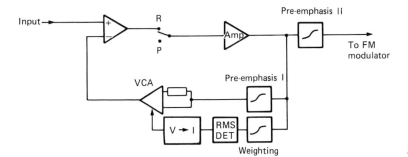

Figure 9.7.   Compressor

frequency range been 2 KHz and 10 KHz, which contains the most energy in speech and music programmes. Extra compression within this range permits more pre-emphasis without the risk of over-deviation. The introduction of a filter within the compander to perform this function is called 'weighting' the frequency response of the compressor.

A block diagram of the recording compressor appears in Figure 9.7. The input signal is applied to an operational amplifier then via a buffer amplifier to the recording pre-emphasis II circuit, whose frequency response is shown in Figure 9.8(b). Compression is performed in the operational amplifier according to the feedback signal applied to its inverting input. A signal sample is fed to a Voltage Controlled Attenuator/Amplifier (VCA) through a filter, Pre-emphasis I. As Figure 9.8(a) shows, this boosts all frequencies above 1 KHz to improve signal to noise ratio. A weighting filter (curve, Figure 9.8c) progressively increases the signal level for increasing frequency in the range 1 to 10 KHz, beyond which a slight roll-off takes place. Thus carefully filtered the signal is presented to the *control* input of the VCA, where as a result higher input frequencies will undergo more compression. The extra compression in the treble range will compensate for the effects of pre-emphasis I and pre-emphasis II in the recording circuits to ensure that FM deviation limits are not exceeded.

In playback mode the frequency-tailored compressor is inserted not in the feedback path of the input amplifier, but in the direct path, as Figure 9.9 shows. The effect of this is to reverse the operation of the circuit, which now becomes an expander whose amplitude- and frequency-response mirrors that introduced during record. All frequencies that were compressed and pre-emphasised are now de-emphasised and expanded

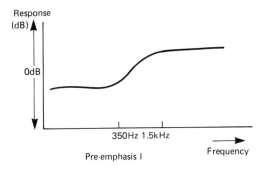

Figure 9.8a.

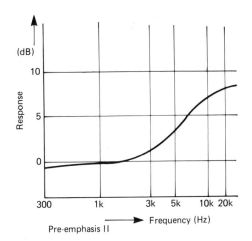

Figure 9.8b.

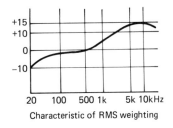

Figure 9.8c.

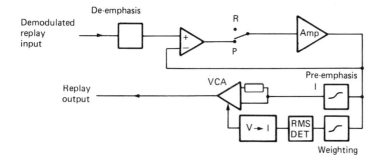

Figure 9.9.   Expander

with the result that tape noise is also attenuated – but over a much wider frequency range than can be achieved by the Dolby system.

RMS detector

An RMS (Root Mean Square) detector is used in the compander to operate the VCA from the 'average' level of the audio input signal. Conventional average or mean level detectors are not good enough for this application where a more accurate representation of the average level of a complex signal is required. A peak detector is not sufficiently accurate either; consider the two signal forms shown in Figure 9.10. It can be seen that neither the peak values nor the mean values are the same for the waveforms (a) and (b), though the

energy content (the shaded area within the waveform envelope) is identical in both cases. This can only be proved by taking the peak value, squaring it, and then taking the square root of the *mean* value of all the 'squared peaks' to produce the effective value:

$$\text{Effective value} = \sqrt{\text{mean value of (peak values)}^2}.$$

Audio head switching

During replay it is necessary to switch between the audio heads, selecting each alternately as it scans the 180° tape wrap, in just the same way as for video heads. The same RF switching (flip-flop) squarewave is used, but because the audio heads are lagging 42° behind the video heads their

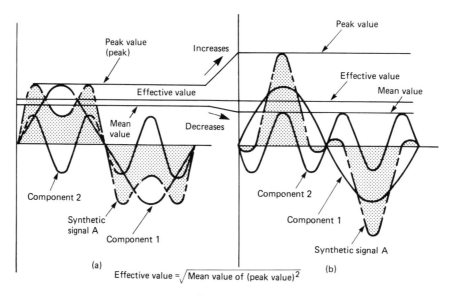

Figure 9.10.

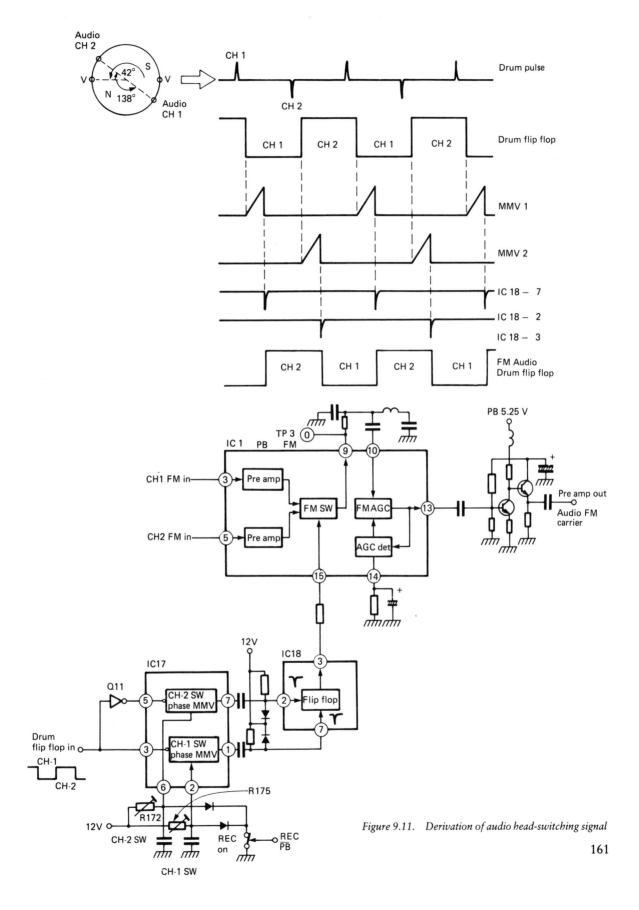

Figure 9.11. Derivation of audio head-switching signal

161

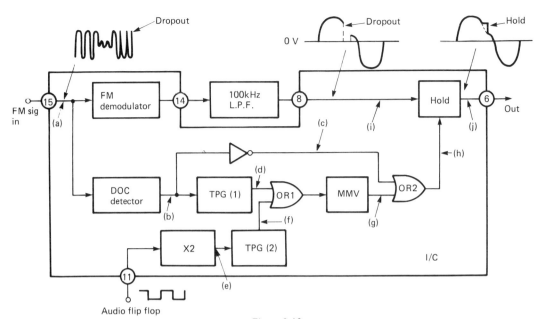

Figure 9.12a.

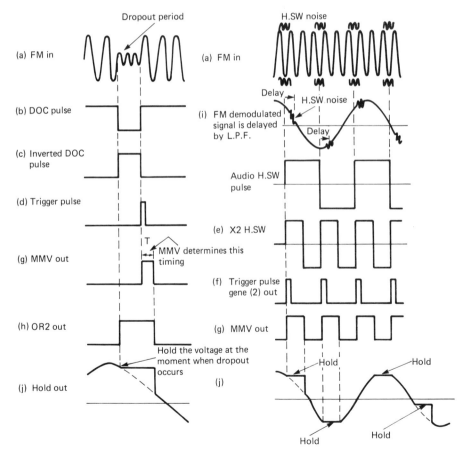

Figure 9.12b.

Figure 9.12c.

162

switching signal must be delayed by 42°, which corresponds to 4.7 ms.

The basic diagram and timing waveforms (JVC HRD725 version) are given in Figure 9.11. The drum flip-flop signal is sent to two 4.7 ms delay monostables in IC17 via direct and inverting paths respectively, so that they are alternately triggered by positive and negative edges of the squarewave. The reset edges of the monostable outputs are sharpened into trigger pulses in a differentiating network, then used to set and reset a bistable. Its output forms the audio head-switching waveform, now delayed by 4.7 ms from the video head switch points, and able to gate the alternating FM-audio heads' outputs into a continuous FM carrier.

Drop out compensation

A momentary loss in FM audio carrier is masked by applying a short 'hold' pulse to the demodulated audio signal to prevent it from falling to zero. This hold level is generated within the IC, and fixed at about 75% peak audio level. It is also inserted at the instant of audio head changeover; the hold pulse is derived from the audio head-switching flip-flop.

When a drop out occurs the result is a reduction in the level of the replay audio FM carrier, possibly to the point where noise or distortion will mar the demodulated audio signal. At the moment of switching between audio heads there is further risk of introducing noise and distortion. In both cases a masking technique must be used to preserve good sound reproduction. For drop out the 'hold' period is extended beyond that of the drop out itself by use of a short monostable.

Figures 9.12(a), (b) and (c) show the basic circuit and waveforms involved. In diagram (a) the FM carrier signal is demodulated then passed through a 100 KHz low pass filter (to remove residual FM carrier) before being applied to the hold circuit. FM carrier is monitored by the drop out detector to generate a drop out pulse which via an invertor provides one input to the OR2 OR gate. A trigger pulse (generated from the trailing edge of the DOC pulse) sets a short-duration monostable via the OR1 OR gate. Monostable output is applied to the second gate of OR2, whose output pulse is thus prolonged; these actions are clarified in the time-related waveforms of Figure 9.12b.

Figures (c) and the lower portion of (a) are concerned with head-switching noise masking. Audio-head switching pulses are frequency-doubled to facilitate the generation of trigger pulses in time-coincidence with both positive and negative edges of the head-switch signal. These trigger the monostable via the lower gate of OR1 so that a 'hold' pulse is generated at the instant of each audio head switchover.

As waveforms (j) in both Figure (c) and Figure (d) show, the hold circuit puts a 'lump' on the audio signal. It can be ironed out by a de-emphasis filter downstream, though JVC differentiate the audio signal and add the resulting spikes back to the hold section in inverted form. This has the effect of softening the edges of the hold pulse and rendering a smooth and harmonic-free output from the following 20 KHz de-emphasis filter.

10

VHS-C and camcorders

The VHS-C format is a small version of the standard VHS format: while cassette size is only about one third of that of a standard cassette the recordings are VHS compatible. It first appeared as a small compact portable video recorder, part of a camera/videorecorder package to meet the demand for a lightweight VCR that could be carried on a shoulder strap comfortably. The small VCR was not as popular as the standard size, it seemed that the requirement for up to 3 hours recording time outweighed the ligher VHS-C recorder with only 30 minutes capability.

The VHS-C format system was adopted by JVC for its range of camera/recorder combinations which became known as the camcorder or videomovie, and provided for up to 30 minutes of recording time. Full size VHS camcorders were also produced by other manufacturers but the cassette housing determined the size of the unit which was not heavy but nevertheless it was bulky and difficult to balance for long periods.

To overcome the criticisms of its camcorders with regard to recording time JVC designed a VHS-C format camcorder with a CCD pickup tube and half-speed long-play facility, model GR-C7. The ability to record for up to an hour with a very lightweight camera/recorder was popular with other Japanese manufacturers and some European manufacturers also. JVC produced clones for many other brand names. Later in 1986 a record-only model was announced which with cassette and battery weighed less than 1 kg.

VHS-C system cassette

The supply spool is a small version (Figure 10.1) of the standard VHS cassette spool, and can sit upon an open turntable within the video recorder. The spool socket is the same as a standard cassette despite the small diameter of the spool. The take up spool is mounted internally on a small spindle and it is not driven by its axial but by a toothed gear on its circumference, as shown in Figure 10.1.

Cassette adapter

When the VHS-C system cassette is mounted within a cassette adapter then the recording can be played on a full size VHS videorecorder. The supply spool sits upon the VCR turntable and the take up spool is driven from the VCR take up turntable via a small intermediate gear (G) arrangement within the adapter (Figure 10.2). Earlier versions of the cassette adapter were 'manual' in that the loading arms A and B had to be moved outwards and forwards to match the tape path, manually, by a side mounted wheel and

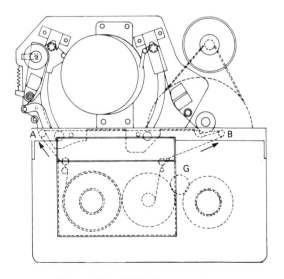

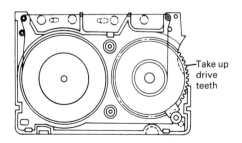

Figure 10.1.   Compact videocassette

Figure 10.2.   VHS-C cassette adapter

internal gears. Later versions use a battery driven motor actuated by the VHS-C cassette being inserted into the adapter. In either type of adapter the tape is pulled out of the VHS-C format cassette to the front of the adapter thus matching it to a standard VHS cassette. A small point worth noting when putting a VHS-C format cassette into an adapter is that a few seconds of the beginning of the VHS-C cassette recording are 'lost' within the adapter as the internal take-up path is increased.

VHS-C format video head drum

The VHS-C camcorder utilises a very small, lightweight, videohead drum assembly and various arrangements have been made to maintain VHS compatibility. It is only two-thirds the size of a standard VHS videohead drum and has four videoheads mounted upon it, these are designated A, A', B and B' and they are used in turn to play or record the video tracks upon the tape. In order to follow the standard VHS signal tracks, Ch.1 (or A) and Ch.2 (or B) (see Figure 4.2 and Chapter 7), use of the four video heads and their switching sequence is determined by the amount of tape wrapped around the drum.

A comparison of the two video head drums is shown looking from the non-recording side is shown in Figure 10.3. Table 10.1 compares specifications. The slanted track is laid down in the VHS format (refer to Figure 1.6) with an azimuth tilt of ±6° between heads. Head A (Ch.1)

has an azimuth of +6° and head B (Ch.2) has an azimuth of −6° and it follows that the tracks can be allocated A and B. From Figure 10.3a in the standard two headed VHS system it can be seen that head A contacts the tape upon its lower edge just as head B is leaving at the upper edge, hence head A will follow an A track. As the video head travels around the drum the tape moves very slowly by comparison in the same direction, the head travels across the tape as a slant track and off the top edge having gone around by 180° plus the overlap margin of about 3° at the start and 3° at the end making a wrap around of 186° in all.

Let us assume the part of the circumference of the 62 mm diameter drum that the head is in contact with the tape is ℓ, and that it is equivalent to the length of the track for the purpose of this comparison. Ignoring the extra wrap around we can say that the length of the path ℓ is:

$$\ell = \pi \times D/2$$

where $D = 62$ mm. Therefore ℓ is about 97.4 mm.

Table 10.1 Comparison between VHS and VHS-C

Specification	VHS	VHS-C
A head azimuth	+6°	+6°
B head azimuth	−6°	−6°
A' head azimuth	—	+6°
B' head azimuth	—	−6°
Drum diameter	62 mm	41 mm
Tape wrap angle	180°	270°
Rotating speed	25 rps	37.5 rps

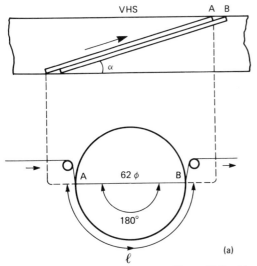

(a)

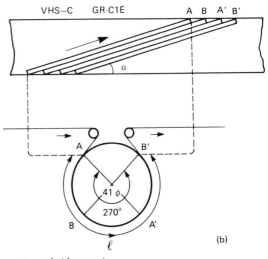

(b)

Figure 10.3. New recording system of video movie

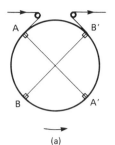 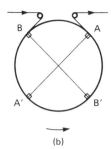 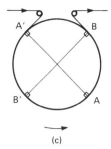 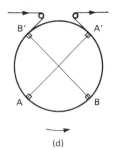

(a) (b) (c) (d)

Figure 10.4. Video head sequence

In the case of the VHS-C format (Figure 10.3b), the video head drum path ℓ has to be the same 97.4 mm for compatibility. As the diameter of the VHS-C head drum is only 41 mm then the total circumference is πD, which is 128.8 mm. Path ℓ is 97.4 mm, which represents 97.4/128.8 or 3/4 of the total circumference. So in order to be compatible, the tape has to be wrapped around the 3/4 of the VHS-C format head drum (3/4 × 360°) which is a wrap around of 270°, plus a total overlap of 7° making 277° in all. We have now established that a smaller VHS-C video head drum requires a greater degree of wrap around in order to keep the same length of tape in contact with the drum to record or replay the standard video track length.

Ingenious use is made of the four video heads on the drum, they are mounted in the order A, B, A', B' and record or playback in that order: an easy way to work out the sequence is to remember that whichever head is contacting the tape then the next one to be used is 90° ahead of it. This is shown in Figure 10.4 a, b, c and d. Initially head A contacts the tape and records a track, as it leaves head B contacts the tape, as head B leaves the tape then head A' takes over to record, as A' leaves the tape then head B' records and as head B' leaves the tape head A takes over once more to repeat the sequence. The tracks are recorded in the order A, B, A', B'. Heads A and A' have an azimuth of +6° whilst heads B and B' have an azimuth of −6° therefore the VHS tracks are recorded and replayed in the standard A, B sequence.

Drum speed

When servicing a VHS-C format camcorder there is a trap that even experienced video engineers can easily fall into. The standard VHS recorders have servo timings which are equivalent to field rate,

that is 40 ms periods. However, careful consideration of the VHS-C format head drum will show that its rotational speed is different, resulting in different servo waveforms.

The standard VHS video head drum rotates at a speed of 1500 rev/min to meet the requirement that a TV field is recorded onto a single track in half of a revolution. A TV field has a period of 20 ms, it follows then that the path ℓ is traversed in 20 ms, so a full revolution takes 40 ms, equivalent to 25 rev/sec or, 1500 rev/min.

In the case of the VHS-C format head drum the path ℓ, now 270°, is still traversed in 20 ms to maintain compatibility. If 270° is equal to 20 ms then 360° is 26.6 ms, a rotational speed of 37.5 rev/sec or, 2250 rev/min.

To summarise, the VHS-C format video head drum is VHS compatible, it is 2/3 the size of a standard VHS video head drum, it has four video heads and rotates at a speed which is 50% faster.

Video head switching

In a standard VHS videorecorder the video heads are switched in playback in order to prevent noise pick up on the video head that is not in contact with the tape, so it is switched off. In record mode switching is not necessary. In the VHS-C format there are usually three video heads in contact with the tape at any given time, but as only one is active, it is essential to switch off the others to prevent any sort of crosstalk between heads.

This is illustrated in Figure 10.5 and shows the position of head A around the circumference of the head drum after it has contacted the tape and travelled along 30% of its track. The shaded portion represents recorded tape and head A has started at the lower edge of the tape and is travelling up and along the centre of its track. The position of the other head at this point in time is determined by the fixed tip position in relation to both angular position around the drum and relative tip height. As you can see from the illustration at this point all four heads are in contact with the tape and as this is a recording situation the need to switch A', B and B' fully off is obvious to prevent A' and B from damaging a previous track recording and B' from altering the A track currently being recorded. On the other hand, in playback mode severe crosstalk would result from head B replaying a B track, although the A' and B' heads would not produce much interference as they are the wrong azimuth for the tracks that they cover. The video head to record after A will be B and by the time it has travelled from its current position to the start of its track then the tape will have moved along correspondingly.

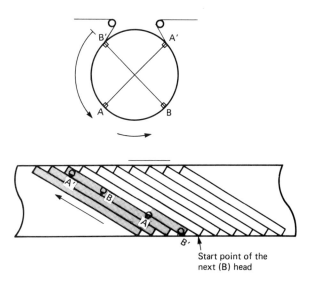

Figure 10.5.

Overlap

In the case of a standard VHS two-head drum, both heads are driven by the FM recording signal and as they are 180° apart while the wrap around is 186° then there is duplicated information at the start and end of each track within which replay head switching occurs to ensure signal continuity and interference-free switch points (refer to Figure 4.2). This duplication period is the overlap region and it is not possible to obtain it by the same method in the VHS-C format because the video heads are switched in record as well as playback. The overlap period must be determined in order to provide for interference-free head switching in playback and it is achieved by the relationship of the head switching signals by switching one head on, slightly before its predecessor is switched off.

In the example in Figure 10.5 when head A nears the end of its track and head B has contacted the tape, a co-existence occurs as the heads are 270° apart and there is a 277° wrap around and so for 3° both heads are contacting the tape. It only requires for the head switching to be modified so that head B is switched on for about 1 ms before head A is switched off. This corresponds to about 16 lines overlap period and as the replay RF switching signal is centred within the overlap period it gives a margin of + or − 8 lines. Figure 10.6 shows the main switching waveforms of 12.5 Hz and 25 Hz. As in the standard VHS videorecorders the 25 Hz signal is the RF switching signal or flip-flop and the 12.5 Hz signal is obtained by division from it. The four head

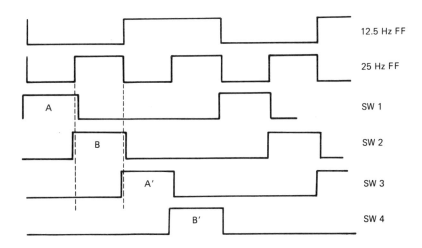

Figure 10.6.

switching signals SW1–SW4 are synchronised to the RF switching 25 Hz but the timing is modified so that each one switches 'on' (high level) 8 lines (518 μs) before the RF transition and 'off' 8 lines after the RF signal transition. The recorded overlap period is therefore centred within a 16 line overlap period leaving sufficient margin of error for interchange tolerances in replay. In practice the timing signals are accurately developed digitally within a single IC and are not dependent upon capacitors or susceptible to drift.

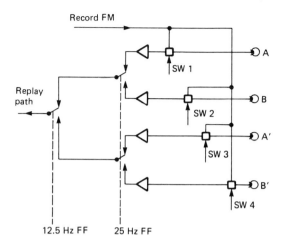

Figure 10.7.

Timing

Figure 10.7 shows a block diagram of the video head record and replay circuits. The recording FM carrier is applied to the video heads via the head switching signals, this ensures that excess signal for the overlap periods is recorded. In replay the head switching signals are still active to prevent crosstalk as previously discussed however the

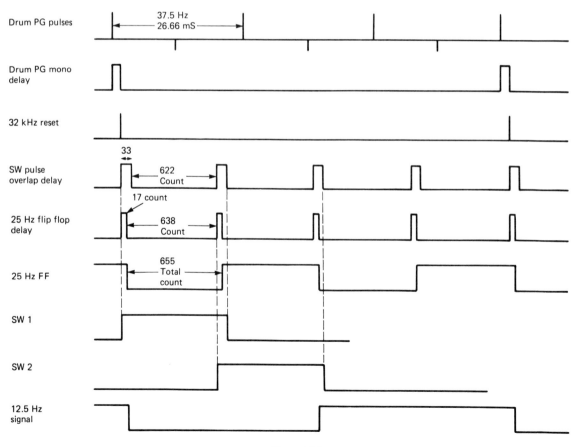

Figure 10.8.

168

replay signal path also embodies the 25 Hz and 12.5 Hz switching signals. As each video head is enabled in turn by the individual head switching signals the precise switching point and video head is selected by the combined logic of the 12.5 Hz and 25 Hz.

The head switching waveforms are derived within the main servo IC and all switching and servo timings are determined by a 32.7 kHz crystal clock. Clock pulses are counted down and all pulse timing determined in a manner similar to the examples given in the section on digital servos in Chapter 4.

The timing diagram is shown in Figure 10.8. Two inputs are used, a 32.768 kHz clock with a pulse period of 30.53 µs and the drum pick up PG pulses generated once each revolution of the drum cylinder. As the drum PG pulses are at a frequency of 37.5 Hz they are divided by 3 to 12.5 Hz or 80 ms. The pulse developed from the drum PG monostable is about 1.9 ms and is variable. As it provides a delay between the drum PG and the start of the 25 Hz flip-flop it can be used to set the replay switch phase adjustment to set the RF switching point to 6.5 lines prior to field sync, this adjustment is now restricted to a single adjustment (PG1) and not two (PG1 and PG2) as in standard machines as the rest of the timing is digital. Note that it takes three full revolutions of the drum cylinder to play or record with all four video heads.

Upon the falling edge of the PG pulse the 32 kHz oscillator is reset and a single clock pulse is generated, this is the preset pulse. The preset pulse resets all counters and initiates the start of the 80 ms timing period. Two counter networks are responsible for the overlap pulse delay and the 25 Hz delay pulse trains. The overlap pulse is a count of 33 clock pulses high and 622 clock pulses low and four of these occur between each preset pulse and it can be seen from the diagram that the 33 count is in fact the overlap period. The 25 Hz flip-flop delay signal is high for a count of 17 clock pulses and low for 638 clock pulses and there are four of these between each preset pulse. In both cases the total count is 655 where 655×30.53 µs $= 20$ ms. In order to time the overlap period each video head switching pulse (SW1-SW4) is set by the rising edge of the overlap pulse and reset by the falling edge, this extends each head switching pulse to 688 counts or 21 ms. The 25 Hz delay period is 17 counts and the 25 Hz RF switching signal corresponds to the falling edge of this pulse, the effect is to put the RF switching pulse

transitions precisely between the overlap pulses with the 12.5 Hz switching signal developed from it.

JVC Videomovie

The Videomovie consists of two main units, the camera and the videorecorder. As the two units are contained within the same housing then some simplification and shared facilities allow a reduced component count as well as lower power consumption, compared to separately housed equipment. For example, supplies are shared and luminance and chrominance signals are maintained separate prior to recording. Figure 10.9 gives an overall block diagram of the camera and recorder sections. Signals from the ½-inch High Band Saticon pickup tube are decoded and processed to PAL colour and luminance, and passed to the recording circuits.

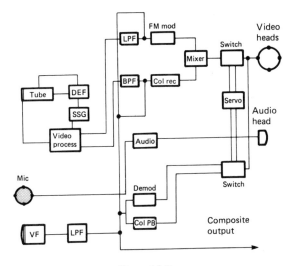

Figure 10.9.

All timing signals are derived from the sync signal generator (SSG) and these drive the tube deflection circuits and provide the very high voltages required for correct tube bias. Syncs, clamping and colour carrier signals are fed to the video process circuits to decode the tube signals and produce separate luminance and chrominance signals. Luminance is passed through a low pass filter and then a dynamic aperture corrector (DYAC) before being frequency modulated for recording. Chrominance is passed through a band pass filter (BPF), not so much for filtering as it is already at 4.43 MHz but for timing purposes as a small delay to match the luminance signal timing during record. Down conversion and colour level control are as standard VHS recording. Luminance FM recording carrier and low frequency colour signals are mixed for recording and then switched to the video heads by the matrix switching circuits.

Simultaneously, luminance and chrominance signals are mixed to a composite colour signal and fed to the output for E/E monitoring. A monochrome viewfinder monitor allows monitoring via a low pass filter (to prevent colour dot patterning).

In replay mode the signals from the video heads are switched again on playback from each video head, the luminance FM signal is limited and demodulated while the colour signal is upconverted and phase corrected, and the two are mixed to a composite signal for the output.

Audio recording and playback are fairly standard and recording is only available from the inbuilt microphone and can be monitored from the main output socket or by earphone.

Camera tube

Light passes through the camera zoom lens, colour temperature filter and a crystal filter focussing the image onto the Saticon tube faceplate and photosensitive target area. Within the tube an electromagnetically deflected electron beam scans the inside of the target area and conforms to the specific TV standard, 625 lines 50 Hz. Variations in light intensity upon the target cause variations in the electron beam's current as it scans the target. Without any light at all some beam current still flows, although it is very small – in the range of picoamps – and is referred to as the dark current.

The tube's target is a high impedance source and the small signal current is very susceptible to external interference. The pre-amp is therefore mounted upon the tube within a screened container embodying the target connection. The target itself is a complex sandwich layer of Selenium-Arsenic-Tellurium (SATicon) film, bonded to other semiconductor layers in a sandwich. A structural diagram is shown in Figure 10.11.

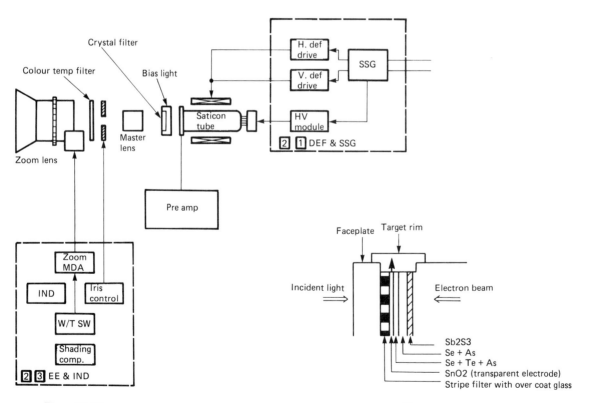

Figure 10.10.

Figure 10.11.

Tube construction is quite complex. Diagonal cyan and yellow stripes are deposited on the inside of a faceplate and overlayered with glass. A transparent layer of stannic oxide (SnO_2), which is the target output electrode, forms a hetero-junction (cross coupled) with selenium-arsenic-tellurium and chalcognide glass material. Between this Se-As-Te layer and a layer of antimony trisulphide (Sb_2S_3) on the beam side is a thick layer of a selenium arsenic mixture.

Light passes through the faceplate, stannic oxide and SAT layers to the selenium arsenic layer where electron and hole charge carriers are produced. Electrons flow towards the faceplate and into the stannic oxide via the SAT layer. Holes, on the other hand, cannot penetrate into the stannic oxide layer due to the selenium film deposited upon it by the SAT layer. The holes travel the opposite way and couple with the scanning beam electrons through the antimony trisulphide. Electrons from the scanning beam cannot pass through the antimony trisulphide due to a film of selenium on the opposite side. The scanning beam current is therefore dependent on the quantity of holes available for coupling.

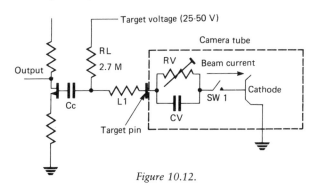

Figure 10.12.

So much for theory, but a much more practical approach can also be realised. An equivalent circuit representing a single pixel element of the photosensitive target layer is shown in Figure 10.12. The Saticon target layer effectively forms a parallel circuit RV and CV. RV is varied by the intensity of the image light falling upon it, with no light RV has a finite value and is designed to be as high as possible to minimise the standing dark current. As the target is scanned by the electron beam each pixel is hit by the beam once every frame. Contact of the electron beam on the target pixel is represented by the switch SW1. When the electron beam hits the pixel, SW1 closes and CV is

charged up between the target and the cathode as the electron beam effectively connects one end of CV to the cathode. After the electron beam has passed CV begins to discharge via RV whose value is light dependent. If there is no light RV is a high resistance and the amount of discharge from CV is very small, if the light level is high then RV is small in value and the discharge of CV is high. The next time the electron beam passes over the pixel the amount of beam current that flows through the target load resistor RL will be the amount required to recharge CV, hence if the light level is high then the charge current through RL is high and the signal voltage across RL increases.

The target output signal is therefore dependent upon the refresh charging current of each pixel capacitor CV and the light variable resistance of each pixel resistance RV. As the Saticon electron beam scans a TV line, many hundreds of small capacitive elements charge up, the variation of charging currents through the target load resistor, RL, develop the signal voltage which is connected to the camera pre amp via Cc.

The signal voltage across RL is very small but it has a very good signal to noise ratio, and this must be maintained. A very low noise junction field effect transistor (JFET) is used for the first stage and is selected for its low noise characteristics. L1 is called a Percival coil, it is used to resonate with the capacitive output of the target and lift the output at high frequencies.

Tube construction

At the faceplate end of the tube, shown in Figure 10.13, is the external connection of the target, the target flange or rim. A small leaf spring contacts the target rim and extends into the pre-amp to connect by soldering to the Percival coil.

One danger when servicing cameras is the presence of very high voltages on innocent looking printed circuit boards. While a shock may not be lethal it can cause the demise of the camera if it is dropped. Some reasonably high voltages are found around the camera tube.

The highest voltage is the mesh voltage of G6 and G3 – around 1,400–1,600 volts. Grid G5 is a collimator lens to narrow the beam as it approaches the target and is at 700 V, functioning in conjunction with G4 which is variable between 200–300 V as the electrostatic focus voltage. If the beam is not focussed the tube resolution drops and the R/B colour carrier reduces to zero leaving a green picture.

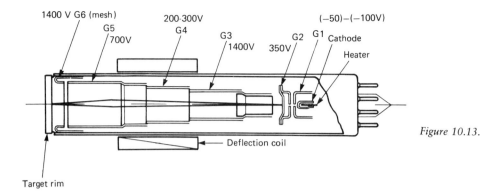

1400 V G6 (mesh)
G5 700V
200-300V G4
G3 1400V
G2 350V
G1
Cathode
Heater
(−50)−(−100V)

Deflection coil

Target rim

Figure 10.13.

G2 is the first high voltage electrode to accelerate the electrons leaving the cathode. G1 is a reverse biased grid and it is variable to set the beam current to a value sufficient to charge CV at full white, without saturating the target. The visual effect of reducing the beam current below the nominal value is to clip the peak whites and it is set to a level just above peak white maximum voltage at the signal output. Too much beam current will adversely affect focussing.

Cathode DC standing voltage is about 6V above chassis, but the tube needs to be turned off during field and line blanking, so blanking pulses are applied to the cathode at about 64V p/p. An oscilloscope on the beam current grid, G1, will show negative blanking pulses down to −70V with a black level of about −50V upon which the video signal can be seen. This measurement indicates a functioning tube even if the output pre-amp has failed, and no apparent output signal can be detected.

While focussing is electrostatic, deflection is electromagnetic and is achieved by a deflection yoke around the tube. Shorted turns within the yoke can stop deflection resulting in no output from the target and no grid signal.

Bias light

To reduce lag or after-image effects caused by a too small dark current, a bias light illuminates the target. If, say, there is a rapid change from bright light to dark it is possible for CV elements to be left with residual charges which could not be discharged through RV between electron beam scans, as RV is too high in value. These charges show up as an after-image. The bias light is used to maintain lower values of RV and hence increase the discharge of the CV elements.

Crystal filter

The crystal filter is used to reduce cross colour effects or moiré patterning from scenes with small chequered patterns or fine stripes. If, within a television, luminance signals of fine patterns around the 4.433 MHz chrominance subcarrier frequency occur, then random blue/purple moiré or herringbone patterns are seen. A similar effect occurs in a camera when scenes which create luminance frequencies around the R/B carrier frequency of 3.9 MHz are viewed. The crystal filter is an optical spatial filter with optical gratings effective at the 3.9 MHz frequency which effectively block fine patterns at the filter.

Colour signal

A colour television tube has coloured phosphors of red, blue and green to enable it to display colours; a colour camera tube has coloured stripes upon its faceplate in order for it to distinguish colours.

Although the target area covers most of the tube faceplate the main area of use is a rectangle of 5.6 by 7.2 mm wherein the purity of material and uniformity of sensitivity are kept within fine limits (Figure 10.14). The image from the lens is focussed onto this area and in order to suppress signals outside this area the rest of the tube is masked off by a black mask. A precise fine stripe filter covers the image area to provide for colour recognition. The main tube signal is green, stripes on the target face are red and blue so within the decoding system white can be obtained. As the light transmission of red and blue filters is low the reverse colours of yellow and cyan are used, cyan being −R and yellow being −B. The stripes cover the target faceplate in diagonals at an angle of 25.40° with a width of 13.95 μm, a gap of

172

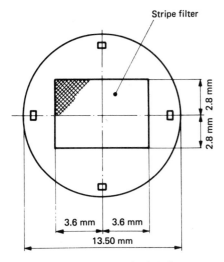

Figure 10.14. Tube faceplate

Stripe filter

2.8 mm
2.8 mm
2.8 mm

3.6 mm | 3.6 mm

13.50 mm

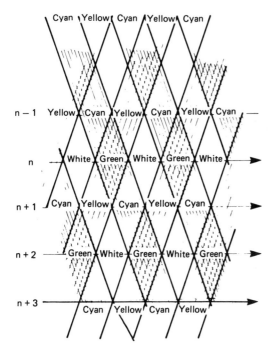

Cyan Yellow Cyan Yellow Cyan

n − 1   Yellow Cyan Yellow Cyan Yellow Cyan

n   White Green White Green White

n + 1   Cyan Yellow Cyan Yellow Cyan

n + 2   Green White Green White Green

n + 3   Cyan Yellow Cyan Yellow

Figure 10.15.

13.95 µm and a pitch of 30.89 µm as shown in Figure 10.15, a magnified view of a small section of the stripe filter. The stripes act not only as colour filters but the physical dimensions and spacing are utilised to generate specific signal parameters. As the electron scanning beam passes over the stripes it generates one output per stripe and as the number of stripes along the horizontal

is fixed then a carrier signal is generated: an analogy is that of a child running along some railings with a stick to generate a tone. In the Videomovie the number of stripes in a line period generates a carrier of 3.9 MHz, it can be noted then that the horizontal width is critical. The green signal is present in all three stripe colours, white, cyan and yellow, and so the target video signal tends to be green rather than white. It is complemented, however, by the 3.9 MHz carrier which contains the red and blue colour information as an amplitude modulation but 180° apart in phase.

Colour separation

In Figure 10.15 the stripes are shown for line scans n, n−1, n+1 and n+2 the relative tube outputs are illustrated in Figure 10.16. As the TV lines are passed through a 1 line delay line we can relate signals from current line n to those of the previous line, n−1. If we take line n and look at the output as it passes over the stripes we get white/green/white etc. As white is equivalent to R+B+G and green is just G we get R+B+G/G/R+B+G and the electrical signal is as shown in Figure 10.16a.

The previous line, n−1, crosses yellow/cyan/yellow, which can be considered to be R+G/B+G/R+G as shown in 10.16b, because the camera faceplate stripes are as in the colour bars on a TV display where yellow is made up of red and green, and cyan is made up of blue and green.

In order to separate the red and blue components, line n is passed through two phase delay networks as shown in Figure 10.17 to give a 90° phase delay Figure 10.16c and a 270° phase delay Figure 10.16d. By using Figure 10.16b as the reference or delay line output n−1, we can subtract from it n delayed by 90° to give the red output Figure 10.16e, and n delayed by 270° to give the blue output Figure 10.16f. The green output is taken as Y and a PAL encoder is then used to provide the composite colour signal.

The tube output is passed through a 3.9 MHz trap and the output green is converted to YH as the luminance signal. A further modification of the green signal by passing through a LPF at 0.9 MHz is YL and is matrixed to R-YL and B-YL. Green will not be recovered until the TV colour decoder and is dependent upon the level of Y with respect to R-Y and B-Y. If light level is too low and R and B signals are lost then the values of R-Y and B-Y will fall leaving the TV decoder with a majority output on the green channel.

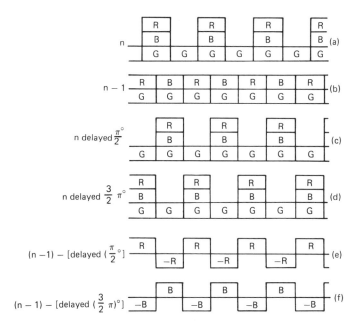

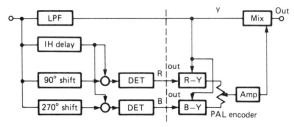

Figure 10.16. *Stripe filter output*

A later development for the JVC camcorders simplified the colour separation circuit, shown in the block diagram of Figure 10.19. First, green is eliminated by passing the R/B signal (a 3.9 MHz carrier) through a band pass filter. The R/B signal (n−1 in Figure 10.16b), can be split into the two components of red and blue 180° out of phase as (R−B). Signal n in Figure 10.16a, however, comprises red and blue signals in phase (R+B).

This can be confirmed by inspection of the stripe filter and comparing lines n−1 and n+1. The yellow stripe has advanced nearer to the start of scan, whereas the cyan stripe has retarded further away from the start of scan. The position of the yellow diamond in line n+3 is the same as n−1 so that the pattern repeats over four lines, so it is not difficult to work out in terms of phase that the stripes shift 360°/4, or 90°/line and that the yellow

Figure 10.17.

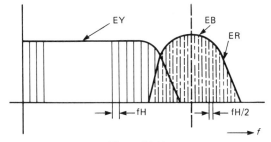

Figure 10.18.

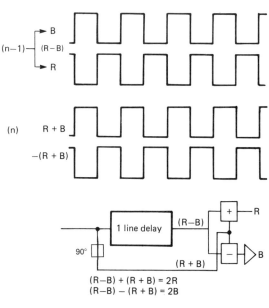

Figure 10.19.

174

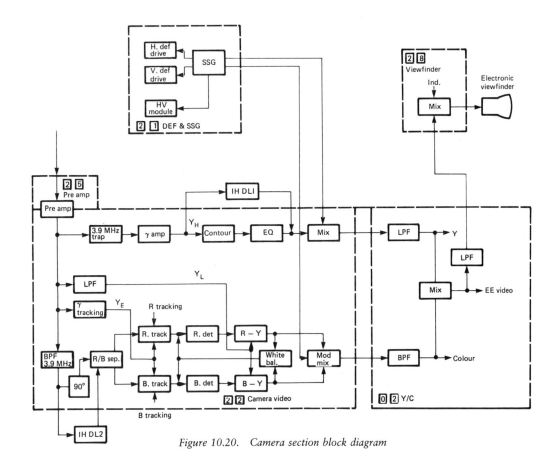

Figure 10.20. Camera section block diagram

phase advances whilst the cyan retards. In terms of red and blue then we have a constant 180° phase shift between them which is rotating at 90°/line. It therefore becomes possible by the use of a 90° phase shift and a 1-line delay with a sum and difference network to separate red and blue.

We can now say the signal coming out of the delay line in Figure 10.19 is the difference signal $(R-B)$ while the direct signal, brought into line with a 90° phase shift is the sum signal $(R+B)$. Separation in a sum network is $(R-B)+(R+B)=2R$ whilst in the difference network we have $(R-B)-(R+B)=2B$, where the negative sign indicates a 180° phase shift which is easily accommodated within the network by a unity gain inverting amplifier to give 2B.

Figure 10.20 illustrates the overall colour video signal handling. In better quality broadcast cameras there is an aperture correction circuit to crispen up the pictures by correcting transition edges in both the horizontal and vertical directions. In the VHS camcorder there is a vertical

edge enhancement circuit which by the subtraction of direct and delayed TV lines will enhance the edges, often referred to as contour correction. Horizontal aperture correction is also carried out in the videorecorder record circuits and an explanation can be found in the section on dynamic aperture correction in Chapter 3. Also, the colour signals undergo tracking and white balance correction.

Gamma correction and tracking

In a single tube camera the levels of the Y signal and the R/B carrier do not remain constant with respect to each other throughout the range of input light. At low light levels the R/B carrier is lower in proportion to the Y video output than at higher levels. This cannot be allowed in high sensitivity colour cameras as there would be a tendency towards dark green pictures as the red and blue outputs fall off if the camera is used in lower light levels. It can be seen in Figure 10.21 that the blue sensitivity is lowest and that red falls between it and the higher Y level. In mid-range light levels the differences remain constant, but at the extremes of low and high light levels the red and blue signal levels fall off first.

It is necessary to ensure that the gains of the red, blue and Y (green) signals remain constant with respect to each other throughout the light range to prevent different tints at different light levels. To this end a triple level tracking correction system is employed as shown in Figure 10.22.

Tracking is corrected by adjusting the gains of the red and blue control amplifiers against the Y reference signal which has already been gamma corrected against tube gain non-linearity. Tracking (1) works in the range up to 40%, tracking (2)

works between 40% and 80% and tracking (3) operates at high light levels only.

The system used is similar in most camcorders and is incorporated within two ICs, shown in Figure 10.23. IC1 is referred to as the gamma tracking generator and produces three signals derived from the luminance input. Pin 16 produces an output which is clipped at 40%, hence any luminance signals up to 40% peak white level are sent to the red and blue tracking (1) controls to white balance the picture up to this level. Pin 17 produces an output of luminance signals only between 40% and 80% to control the white balance between the two levels. Pin 15 outputs peak white signals only to white balance the peak white levels. It is not easy to adjust these controls without the correct lighting conditions of 3,200K

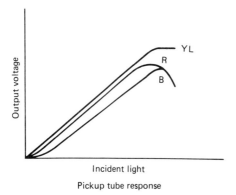

Figure 10.21.   Pickup tube response

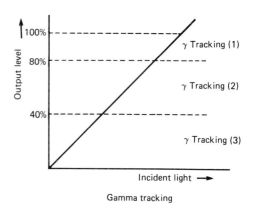

Figure 10.22.   Gamma tracking

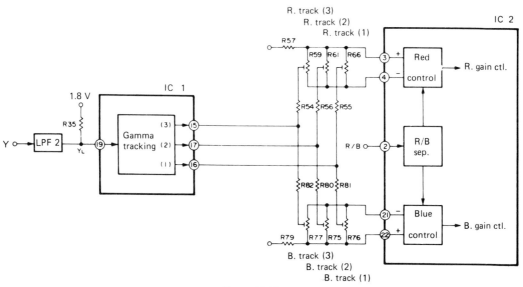

Figure 10.23.

176

and a grey scale chart. They interact at the crossover levels of 40% and 80% and considerable interadjustment is required to get the whole grey scale tracking correctly, and usually some compromise is found. The peak white levels can only be adjusted if control over the automatic iris is possible or a small white area is illuminated within a black background to fool the iris control and open up the isis.

CCD photo-sensitive pickup device

The solid state photo-sensitive pickup device used in colour cameras to replace the saticon tube is more colloquially known as the charge coupled device (CCD) pickup or imager; a term which refers to the fact that shift registers within the device operate on the charge coupled principle. It is often referred to as a 'tube' although it is a flat cellular array. Advantages are its compactness, low power consumption, resilience to burning and long life. Unlike a camera tube there is no heater element and as the photo elements are solid state there are no adjustments for installation, temperature variation, for ageing or usage. There are two main types of CCD unit; frame transfer and line transfer. In a frame transfer CCD the entire picture is transferred from the sensing area to a storage area, as shown in Figure 10.24, and then read out in a manner to be described for the line

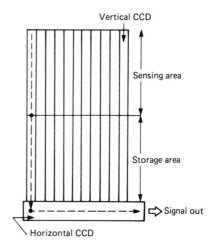

Figure 10.24.

transfer type. Although the frame transfer CCD is fine for monochrome it is not so suitable for colour.

Line transfer CCD

This is the type used in colour portable cameras and camcorders. A CCD solid state pickup element is a large scale integrated circuit consisting of about 200,000 photo diodes forming picture elements, or pixels, on a 7 mm square substrate along with more than 450 shift registers and accompanying pulse control circuits.

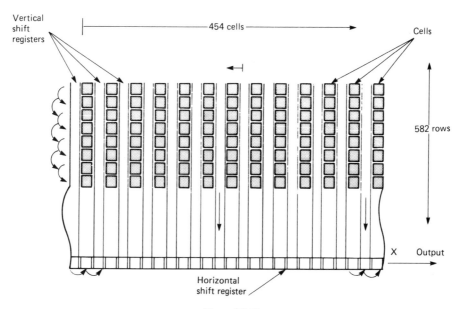

Figure 10.25.

The cell array is 454 horizontal by 582 vertical, that is 582 rows of 454 pixels. In Figure 10.25 the picture elements are shaded, to the left is a vertical shift register element which forms part of a shift register column.

During vertical blanking the charges accumulated on all of the individual photo diodes according to incident light levels are shifted simultaneously left onto vertical shift registers. The vertical shift registers then operate in tandem during line blanking to shift these charges downwards, a row at a time, until a row reaches the lowest level where it is transferred to a horizontal shift register. Cell charges are then shifted out of the horizontal shift register, one at a time, to provide the serial video output.

The scene which is recorded by the camera is inverted through the lens and when it is displayed on the surface of the CCD it is an inverted mirror image, as shown in Figure 10.26. Scanning of the pixels in the CCD starts at the lower right hand corner and finishes at the top left hand corner. This explains the horizontal shift register at the bottom of the stack: it is there so the first pixel to be read out is that one at the lower right hand corner.

Charges are developed in the photo diode in Figure 10.27 which are proportional to light input and are transferred to the vertical shift register (VSR) via a MOS transistor. Each cell has a transfer gate and to transfer the charge from the photo diode to the shift register a pulse TG is applied to the MOS transistor. Alternate rows are used separately to create odd and even fields.

As each row is transferred down the stack to the bottom during line blanking preceding rows are read out of the horizontal shift register and as each charge corresponds to an individual pixel the read out must be at a very high frequency. As the horizontal scanning steps out about 459 charges in

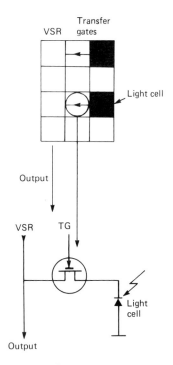

Figure 10.27.

$64\,\mu s$ the stepping frequency will be around $8\,MHz$.

In a similar manner to a camera tube faceplate, not all of the available CCD area is used for the visual picture. An overview is shown in Figure 10.28. The total number of elements in the CCD surface is 454 by 582, a total of 264,228, but the picture area is 418 by 575, a total of 245,525. The rest of the unused cells are maintained at an optically black level by clamping.

Colour

The colour system is very complex using a secondary colour filter with magenta, cyan, green and yellow filter elements instead of just red and blue. Operation is, however, similar to that described for the camera tube where the colour R/B carrier is produced with red and blue components on alternate lines. Figure 10.29 shows the layout of the colour filter overlayed over the pixels. The starting point is the bottom right corner (1, 1) as the first pixel. Note that each row of picture elements is used twice: once in the first field and once in the second field. Use of pixel rows like this may be termed primary and

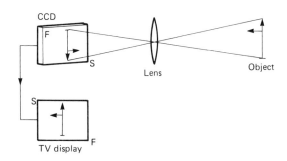

Figure 10.26.

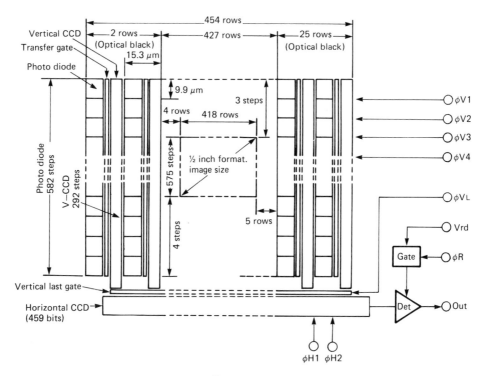

Figure 10.28.

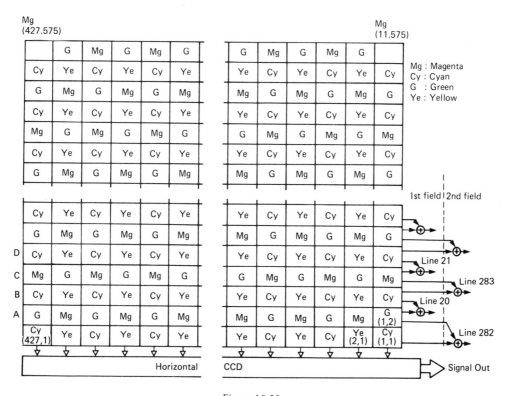

Figure 10.29.

179

secondary; where primary refers to use as the main component and secondary to use as an addition signal.

In the first field, row A forms the primary row of line 20 which, after stepping through the horizontal CCD shift register, outputs G, Mg, G, Mg . . . . . The next row, B, forms the secondary row and outputs Cy, Ye, Cy, Ye . . . . . So, line 20 is a combination of the two colour rows, A and B, such that:

Line 20 = (G+Cy), (Mg+Ye), (G+Cy), (Mg+Ye) . . . . . .

Line 21 is formed by row C as the primary, with colour signals: Mg, G, Mg, G . . . . ., followed by row D as the secondary, with colour signals: Cy, Ye, Cy, Ye . . . . ., such that:

Line 21 = (Mg+Cy), (G+Ye), (Mg+Cy), (G+Ye) . . . . . .

A pattern is set up, shown in Figure 10.30, where each colour line is a combination of two rows of pixels and the pattern is repeated every two TV lines.

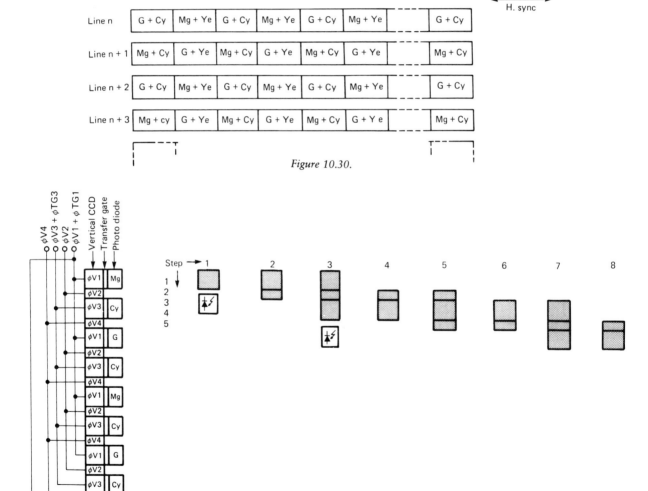

Figure 10.30.

Figure 10.31.

180

Vertical shift register

The shifting of light dependent electrical charges down the vertical shift register stack is carried out with a four phase drive, V1, V2, V3 and V4, with transfer of charge from the photo diode pixels into the shift register by TG pulses attached to V1 and V3 on alternate fields. Figure 10.31 shows the sequence of shifting the charges from a magenta primary pixel and a cyan secondary pixel.

Step 1. V1 goes high and TG1 transfers the charge from a magenta pixel into the first shift register cell.

Step 2. V2 goes high and the magenta charge is shared between the first and second shift register cells.

Step 3. V3 goes high and the magenta charge is further shared out into the third shift register cell. Simultaneously, TG3 transfers the charge from the cyan pixel into the shift register, so the charge stored by the first, second and third shift register cells is a combination of magenta and cyan charges.

Step 4. V1 goes low and both magenta and cyan charges are shared between shift register cells 2 and 3.

Step 5. V4 goes high and the combined charges are shared between shift register cells 2, 3 and 4.

Step 6. V2 goes low and the charges are shared between cells 3 and 4.

Step 7. V1 goes high and the charges are shared between shift register cells 3, 4 and 5. No TG1 pulse occurs in this line.

Step 8. V3 goes low and the charges are shared between shift register cells 4 and 5.

The combined magenta and cyan charges continue down the vertical shift register on this basis as the cycle repeats, until they are shifted into the horizontal shift register.

Horizontal shift register output

The video output of the horizontal shift register is a series of pulses, Figure 10.32, whose amplitudes correspond to the level of charge transferred out of the photo diode pixels. The pulses are not long enough in time to give a video signal and they have to be stretched by the use of a sample and hold circuit to give a continuous video output.

Secondary colour filter

As in the camera Satican tube, use is made of cyan, green, yellow and magenta filters as the light transmission of these colours is better than red and blue, giving improved sensitivity to light. The following explanation refers to Figure 10.33. Cyan filters block red light but not blue or green. Magenta filters block green but not red or blue. Similarly, yellow filters block blue but not red or green.

Now, from the combination of (G+Cy) we get:
(G) + (B+G) = 2G+B.
Also, from (Mg+Ye) we get:
(R+B)+(R+G) = 2R+B+G.

As in the camera example we will produce two signals, luminance and a high frequency carrier containing red and blue.

Each pair of colour signals: (Mg+Ye) and (G+Cy) in Figure 10.33 Ⓐ; and (Mg+Cy) and (G+Ye) in Figure 10.33 Ⓔ are equivalent to the positive and negative cycles, respectively, of a 4 MHz sinusoidal wave, $\omega_c t$.

So, line n (Figure 10.33 Ⓐ = $Y_n + C_n \sin(\omega_c t)$.

And, line n+1 (Figure 10.33 Ⓔ = $Y_{n+1} + C_{n+1} \sin(\omega_c t)$.

Therefore:
$$Y_n = (Mg+Ye) + (G+Cy)$$
$$= (R+B+R+G) + (G+G+B)$$
$$= 2R + 3G + 2B$$
This is a luminance component.

And:
$$Y_{n+1} = (Mg+Cy) + (G+Ye)$$
$$= (R+B+G+B) + (G+R+G)$$
$$= 2R + 3G + 2B$$
also, a luminance component.

Now:
$$C_n = (Mg+Ye) - (G+Cy)$$
$$= (R+B+R+G) - (G+G+B)$$
$$= 2R - G$$
which is an R-Y component.

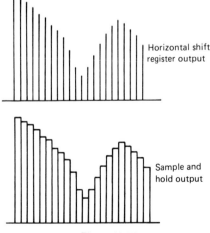

Figure 10.32.

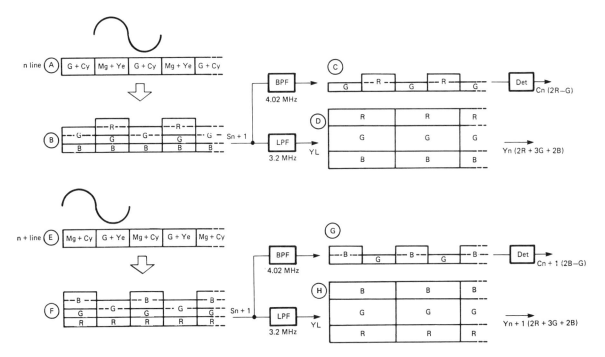

Figure 10.33.

And:
$$Cn+1 = (Mg+Cy) - (G+Ye)$$
$$= (R+B+G+B) - (G+R+G)$$
$$= 2B-G$$
which is a B-Y component.

Line sequential colour difference

From the above we have determined that line n from the CCD matrix contains both luminance (in the form of $2R+3G+2B$) *and* chrominance (in the form of $2R-G$ as an R-Y component). The following line, n+1, also has luminance (in the form of $2R+3G+2B$) and chrominance (in the form of $2B-G$, which is a B-Y component). Luminance is present at all times and can therefore be used to create YH, the full band luminance signal for the recorder merely by passing it through a low pass filter with a cut off at 3.2 MHz.

As was considered for the camera tube we have red and blue component signals in the form of $2R-G$ and $2B-G$, but they are on sequential lines. This is not much good for a PAL system as we require *both* red and blue information for *each*

line. The colour signals are converted to continuous outputs in a system which is an R/B separator, but achieved on a line sequential basis.

The line sequential separator

Figure 10.34 shows the line sequential separator. R-Y and B-Y signals occur on a line-by-line sequence so a single 64 µs delay circuit is used. When an R-Y component enters the delay circuit, a previous line containing a B-Y component is present at the delay circuit output. Two synchronous switches S2 are driven at half-line rate (7.8 kHz) so that they are high for a line period and then low for the next line. In the example shown the upper switch transfers R-Y from the direct signal and the lower switch transfers the B-Y delayed signal to the B-Y output. For the next line the switches change over and the B-Y signal is direct whereas the R-Y signal is the previous one delayed. It can be seen that the two R-Y signals are the same, firstly direct and then delayed, which means that the delayed information is not a true representation of the colour at the actual time the information is used. Nevertheless, we can get

182

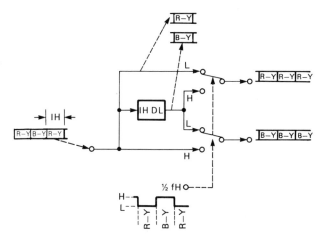

Figure 10.34. Colour difference converter

away with this on the basis that the colour information does not change very much between two adjacent TV lines.

Overall colour system

The overall block diagram of the colour system is shown in Figure 10.35, the output from the CCD consists of the luminance component and the line sequential colour component at 4.02 MHz.

The main luminance signal is YH and has a bandwidth of 3.2 MHz. It is gamma corrected and contour corrected prior to the recorder section. A low frequency variation of the luminance signal,

YL which has a bandwidth of 0.8 MHz to match the chroma bandwidth, is used to eliminate the G component of the 2R−G/2B−G to give 2R/2B. After gamma correction YL is subtracted from the R/B to give R-YL and B-YL which is convenient for PAL/NTSC encoding.

Switch S1 selects the R-YL from the Rch and then B-YL from the Bch on an alternate basis. The line sequential separation then separates R-YL and B-YL.

Automatic focussing

There are a number of quite different techniques for deriving an automatic focus (AF) system, ensuring that the image or subject is focussed onto the saticon, or CCD, faceplate. Some are based on measuring the distance between the camera and the subject, others use the crispness of the image on the photo-sensitive pick-up tube. There are advantages and disadvantages to all of the systems.

Ultrasonic focussing

This is a very basic system using ultrasonic sound to measure the distance between the camera and the subject and uses techniques based upon marine ultrasonic detection. An ultrasonic beam of a frequency around 40 kHz is directed out from the camera to the target subject and bounces back to a

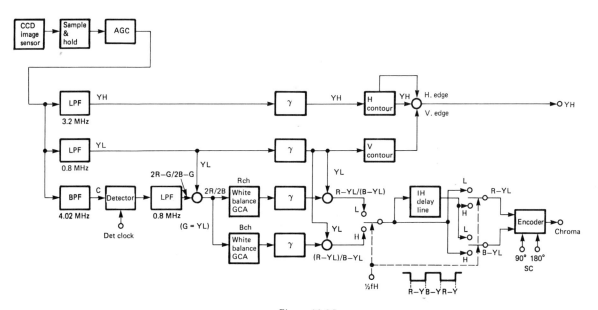

Figure 10.35.

receiver (Figure 10.36). A timer measures the time taken between transmission and reception of the beam in order to determine the distance of the subject and the focus is set accordingly.

It is not a closed loop servo system as the focus point is set indirectly by measurement, and initially the focussing system has to be calibrated, or setup. If drift occurs between the measuring section and the focus drive then it cannot automatically compensate and has to be recalibrated.

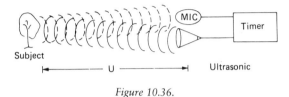

Figure 10.36.

Main limitations include the fact that the ultrasonic system cannot work through a glass window as the glass reflects the beam so the camera focusses upon the glass and not the subject through it. Also, ultrasonic beams can bounce off adjacent objects and can be reflected off at an angle from inclined objects instead of back to the camera making reliable focussing a problem.

Infrared focussing

In an infrared (IR) system the infrared beam is emitted from a light emitting diode and is introduced out of the lens system via a projecting lens mounted above the main lens. It travels out as a beam to the subject image. As the infrared beam travels out of the camera it spreads, although the spread is not very wide, producing a circle about 80 mm diameter some 3 metres from the camera. This is used as the focussing 'spot', hence the

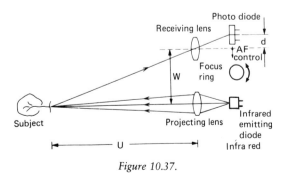

Figure 10.37.

camera will focus upon any object the spot hits, providing that enough infrared light is reflected.

As shown in Figure 10.37, the beam hits the subject and is picked up by a separate receiving lens, normally mounted beneath the main camera lens and focussed upon a receiving IR photo diode. The distance W is fixed between the IR projecting and receiving lenses so that U can be calculated by trigonometry. In practice the distance d between the optical centres of the receiving lens and photo diode is used, as the photo diode is attached to the focus lens ring. When the focus ring rotates, driven by the motor, the AF control moves the photo diode assembly up or down increasing or reducing d, until maximum IR illumination of the photo diode is obtained. If the camera is off focus then the photo diode will receive less illumination as the received beam will not be on the centre of the diode.

This autofocus system operates as a closed loop to maximise the infrared beam onto the photo diode. Two photo diodes are actually used, to improve the control range and provide for more accurate focussing of wider lens settings by maintaining even illumination on both photo diodes. Operation is as follows: the infra red beam is modulated and is switched off when not in use, as a check the beam is pulsed on and off at intervals. If focussing is incorrect then the received beam will not produce equal outputs from both photo diodes. The focus motor is operated to move the photo diodes to an equal illumination position. In doing so the focus ring is moved to obtain the correct focus.

The mechanical tie up between the rotating focus ring and lateral movement of the diode receiver is critical for linear operation and must be correctly set up. Setting up is by adjusting the distance d for best focus at 3 metres.

Limitations are similar to ultrasonic focussing, in that there are problems working through glass as the camera will focus upon the glass. The system also suffers from parallax and reflection problems but not to the same extent as ultrasonic methods.

Solid state triangular system

In the SST system, two mirrors and a double prism focus the image onto a CCD detector as shown in Figure 10.38. Distance d is dependent upon distance U; as U decreases d increases, and this measurement is used as the datum for the

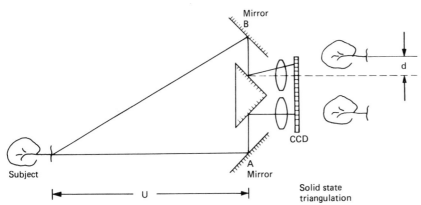

Mirror
B

Subject

U

A
Mirror

CCD

d

Solid state triangulation

Figure 10.38.

autofocus system. Parallax error can be a problem and the physical arrangement is subject to shock hazard.

Frequency detection

As the image of a subject is focussed on the pick-up tube it becomes crisper and sharper, the electrical signal will therefore contain more high frequencies. Autofocus adjustment (Figure 10.39) is used to maximise the high frequency content of the video signal and is therefore a closed loop system. One drawback is that moving subjects create spatial frequencies which are higher than when stationary and can cause the autofocus to hunt backwards and forwards for an elusive mid point.

Through camera lens

Referred to as the TCL system it comes in two forms; both used on Videomovie cameras. The TCL system avoids parallax errors and allows for a greater autofocus range between telephoto and wide angle than other types, while not being affected by glass.

The type of system first used is shown in Figure 10.40, where a portion (30%) of available light input is deflected off via a half prism, mirror and autofocus lens to the TCL sensor. As the autofocus prism cuts down the available light it also reduces the sensitivity of both the camera and the autofocus system. Accuracy is maintained by the fact that the light is split *after* the main lens, so that the TCL sensor obtains the same image as that on the Saticon.

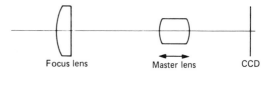

Focus lens

Master lens

CCD

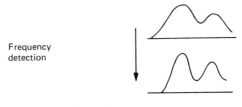

Frequency detection

Figure 10.39.

A second version (Figure 10.41) had the autofocus lens attached to the main focus lens so that the two moved in parallel. This allowed for more light to reach the CCD pick-up tube and the TCL sensor thus increasing the sensitivity of the camera as well as allowing the autofocus to work at lower light levels. Accuracy in this case is maintained by tracking the autofocus lens and the focus lens mechanically.

In both types of autofocus lenses the special TCL sensor (a CCD detector) converts the light into an electrical signal sufficient for a microcomputer to calculate focussing errors and adjust to eliminate them. The principle is similar to that of the split prism used in photographic cameras using two derived images.

Figure 10.42 illustrates the split prism photographic principle whereupon two images are produced within a central circle. They are

185

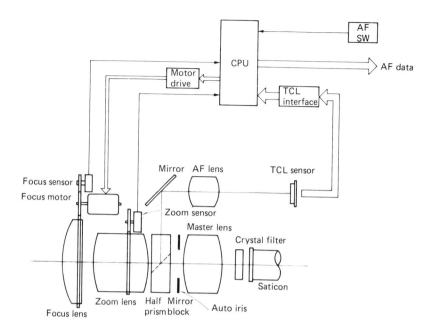

Figure 10.40.

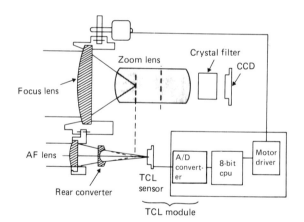

Figure 10.41.

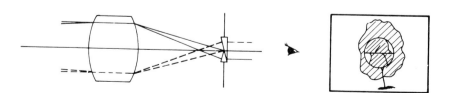

Figure 10.42. Split image prism system

displaced with respect to each other except when the lens system is in focus. A microcomputer can automatically gain the correct focus by sensing the displacement. The two images are produced by two sets of photo diodes 'A' and 'B', mounted beneath a microlens array in a strip as illustrated in Figure 10.43 a and b. CCD registers are used to shift out the charges as two analogue sample signals. The complete CCD image sensor is mounted within an enclosed unit as shown in Figure 10.44, where incident light passes through a protective shield plate and collector lens onto a strip of microlenses covering the pairs of CCD photo diodes.

If the image is not focussed correctly upon the microlens array then the two images A and B produced upon the CCD sensor will be displaced, see Figure 10.45. Only when the lens is focussed will equal amounts of light pass through any microlens and equally upon the A and B photo diodes and produce equal and simultaneous A and B outputs.

A more detailed illustration of the A and B images formed on the CCD sensor is shown in Figure 10.46. In Figure 10.46b the focal plane is formed on top of each microlens and so equal amounts of light fall upon the A and B photo diodes. The two images are coincident and form a unity output.

In Figure 10.46a the lens system is in front of correct focus and the focal plane is well in front of the CCD chip. In cells 1 and 2 light only falls on the A photo diodes and so it puts the A image in front of the B image as the cells are shifted out by the CCD registers.

In Figure 10.46c the lens system is rear of focus and the focal plane falls to the rear of the CCD chip. This time, in cells 1 and 2, light only falls on the B photo diodes and the B image is consequently in front of the A image as the CCD elements are read out.

From this information the microcomputer can deduce front or rear focus and obtain the correct focus by coinciding the two images. Whilst this system works very well, limitations still exist. For example, the system cannot function in low light levels on plain surfaces or in conditions where the charge on the CCD cells is insufficient to allow for focus detection. Even if the subject is bright, such as white wall, there may be insufficient contrast to form an image and the A and B outputs will be a straight horizontal line, coincidence cannot therefore be confirmed and the auto focus system will shut off. Also an image with horizontal lines (say, venetian blinds) will cause the same problem as a white wall while an image with vertical lines (say, fence posts or trees) will confuse the coincidence determination by creating a number of A and B images the same so that displacement cannot be determined.

Initially the sensor formed a sensing rectangle window in the centre of the picture area as shown in Figure 10.47 as used in the JVC GRC2 and equivalents. The later GRC7 had a rectangular focus window mounted at 45° which increased in size when zoomed in (Figure 10.48a) and reduced

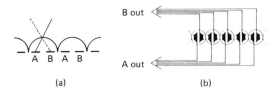

Figure 10.43.

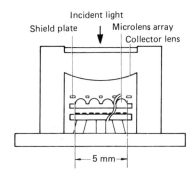

Figure 10.44.

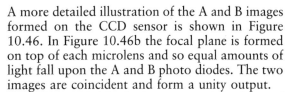

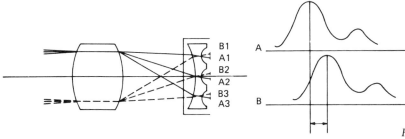

Figure 10.45. TCL image sensor system

187

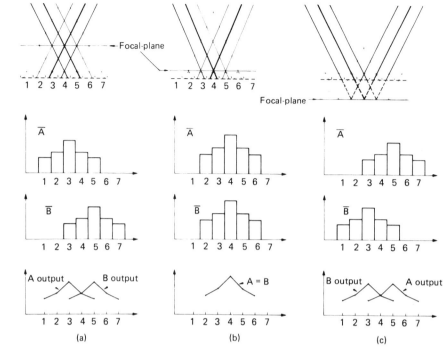

Front focus   Exact focus   Rear focus

Figure 10.46.

(a)   (b)   (c)

at wide angle (Figure 10.48b). This was achieved by the use of the moving AF lens, as illustrated in Figure 10.41. This improved the auto focus performance between wide angle and telephoto.

The autofocus system follows a microcomputer algorithm program:

1. The autofocus/manual switch is checked for auto position, and the AF system switched on.

2. Start CCD photo diodes charging and monitor the charge, setting the average charge to the centre of the A/D converter range.

3. Read out A and B images from CCD cells and convert to digital information.

4. Compute image deviations, if low contrast go back to start.

5. Convert results from split image deviation into focus deviation, i.e. distance from TCL sensor to focal plane. If zero then exact focus has been obtained. Return to start.

6. Convert focus deviation into distance from focal plane to camera subject.

7. Calculate focus ring rotation.

8. Switch on focus drive motor and rotate focus ring by calculated amount. Return to start and switch off motor.

Calculation of the focus ring rotation required and subsequent rotation is confirmed by pulses issued by the focus sensor pulse generator.

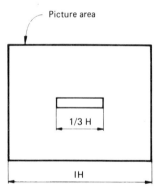

Figure 10.47.

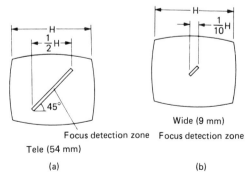

Focus detection zone   Focus detection zone
Tele (54 mm)

(a)   (b)

Figure 10.48.

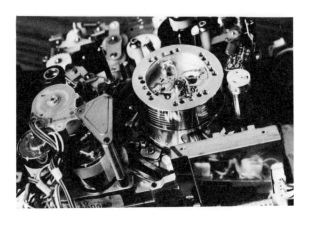

Plates 10. 1–5.   C format mechanical layout showing 40 mm videohead drum and four video heads (JVC GR-C2)

Plate 10. 6.   JVC GR-C2 showing autofocus drive motor (the bit with 5A on the label) and the focus rotation detector

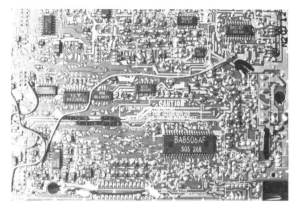

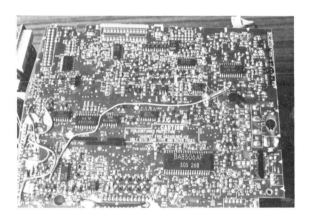

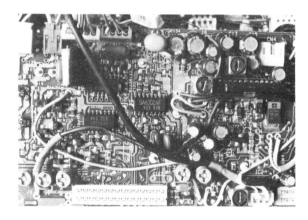

Plates 10. 7–12. JVC GR-C2 camcorder PCBs showing the high density component layout using small surface-mounted components

Plate 10. 13.   Top view of the camera section of a JVC GR-C2 to illustrate the overall length due to the use of a camera tube

Plate 10. 14.   JVC GR-C7 CCD camera. The CCD imager is mounted on the PCB clamped to the end of the lens assembly. To the left is the camera colour processing and video output PCBs

Plate 10. 15.   JVC GR-C7 camera showing the use of LSI I/Cs and connected to the VCR section (in the background) by extension cables. When this photograph was taken the camera was fully operational and producing colour pictures

Plate 10. 16.   JVC GXN70 autofocus camera using an infra red beam. The beam is emitted from the I/R section diode mounted on the top section of the near part of the lens assembly. It is identified by the 'eye' shaped I/R filter lens. The diode is mounted upon an adjustable bracket. Beneath the lens are the zoom and focus motors

11

VIDEORECORDER FAULT PATTERNS

This chapter is a compilation of examples of fault symptoms and causes that have been found in various models of videorecorders. It is not exhaustive, and should be regarded as a guide as to which area of the machine's circuitry or mechanics may be responsible for a given fault symptom. Component number references, where relevant, are given in each case, and these must be related to the service manual for the model involved – there is no room here for detailed individual drawings and circuit diagrams! The suggestions given will not necessarily provide an instant answer to a problem; many widely different faults can give rise to identical symptoms, and nowhere is this more so than in cases of 'no ·deck functions' and 'no colour'. In all fault diagnosis work it is essential to adopt a logical and progressive approach to avoid prolonged and fruitless effort.

A number of videorecorder manufacturers have used a light brown sealant to adhere wires and components to the print side of the PCBs. This can become conductive, possibly hydroscopic, absorbing moisture or trapping moisture beneath it. Whatever the cause, the effect can be faults induced by current leakage to ground or to other components, particularly in circuits with high impedance inputs. If a microprocessor is erratic, or a fault is traced to an area with sealant upon it, then it is worthwhile to remove the sealant. With care, it will come off as a skin-type layer.

To fault find on microprocessor circuits, the best approach is to start by checking the power supply for the correct voltage to the IC and then, with an oscilloscope, look at the reset input and confirm that it is held either high or low for a short period after switch on. Check also the clock pins for high frequency oscillations and then any identifiable pulse output ports, such as key scan outputs, for the presence of pulses. When a halt or fault action occurs check the input ports at the current position for a false signal. For example, a VCR that accepts a cassette, then threads up and unthreads again, perhaps after a short delay of 3–5 seconds has followed its program up to

threading up. At that point it will expect to see that an after loading (AL) switch has contacted, RF switching pulses are present to indicate drum rotation, capstan FG pulses are present for capstan rotation, and take up and/or supply spool rotation pulses are present. If any of the aforementioned do not occur then it is probably for that reason that the emergency unthreading sub-routine has been initiated. It is a logical process although sometimes faults may appear to defy even the laws of physics. Apply lateral thinking, there is a reason, however obscure. Switches, leaky de-coupling capacitors, opto couplers (noisy or leaky) and any type of transducer which can affect voltage levels may be suspect. In the cases of clock and tuning ICs in recorders which have a battery back-up facility – you must disconnect the rechargeable battery before working on the PCB to fit or replace components.

Circuit protectors

Circuit protectors are becoming more common and are generally regarded as a necessary evil by engineers. They are designed to protect sensitive integrated circuits and therefore fail very quickly at their rated current. Circuit protectors replace fuses and have the advantage of being very compact, they come in two forms: a flat transistor (type F) and a D-shaped transistor (type N). The letter is followed by two digits to represent the current rating. F15, for example, is a 600 mA flat transistor, whereas N20 is a 800 mA D-shaped transistor. For whatever reason it was decided upon, the fuse rating is calculated by multiplying the code number by 40 with the result being in mA. It is worthwhile maintaining a workshop stock of F15, F20, N15 and N20.

Akai

VS series:

Intermittent play operation Check that threading is completed and then replace the threading drive belt.

VS1:

No power when function is depressed Check cassette lamp. If this is ok check power supply.

VS4:

No functions After replacing the cassette lamp, if functions are not returned check SF7 circuit protector and ensure the mechanics are reset to STOP before operating. Replace power supply regulator I/C.

VS5:

'Breakdown' indicated Check cassette lamp.
Intermittent or crackling sound Check relay RLI.

Bush – see Toshiba

Ferguson – see JVC

Grundig

General: Always start by checking power rail voltages to ensure that the regulated supplies are to specification.

Grundig 2×4:

Negative picture in E/E mode Check −22 V supply to modulator.

2×4 Super:

No power supply operation Check pulses from IC to switching transistor, check C401 by replacement.
Poor drum servo lock from cold Due to C446/C447/C451 drying up and causing power supply ripples.
System memory data corrupted This can happen for example if the AC power mains are interrupted during a sequence. Reset takes place at the start of unthreading, and can be invoked by hand-winding the motor shaft to set the deck to 'play' position (metal switch segments in contact with wiper) then switching on. All memories now clear to permit normal unthreading.
No power supplies, PSU not oscillating C401 o/c. If the 15 V CL rail on pin 5 of AC connector is missing, pins 5 and 6 can be linked and the accumulator unit removed; there will now be no battery back-up.

Some power rails at 0 V Failure (o/c) of safety resistors R483, R485 or R487 removes DTF tracking control; of R435 also takes out the 33 V tuning line and modulator frequency adjustment.
Failure of accumulator module Faulty zener D2809 and transistor T2809.
No display or power-relay operation Replace IC280.
Head drum runs fast 15 V power rail failure, or no 50 Hz timing signal from DTF micro IC2640.
Tracking errors on replay Missing DTF waveforms; drive transistors faulty or power rail failure.
'Cass' displayed and no action when record selected Similar to result of inserting record-lock protected cassette; crack in print track conveying record-inhibit signal to keyboard micro IC.

Note that there are two versions of IC280 on the keyboard module. Suffix B316 implies the use of two reset subpanels containing IC160 and T170; suffix E316 types do not require the subpanels. Grundig replacement ICs type SDA2010–E316 come with a reset rewiring kit.
Fast forward sluggish Poor take up in playback. Power rail 4 V instead of 12 V due to zener D485 failing.

VS180:

If problems occur in turning the reel motors off even though the logic level FKT is low try replacing C301 and C305.

VS200:

Replay static spots Check wiper springs beneath base plate of the drum motor, mark its peripheral edge to ensure correct re-assembly without any rotational error.
Intermittent stopping during play or record May be due to the pinch roller bearings seizing.
Starts to thread up but only partially takes the tape before stopping Remove plastic dust cover from brake solenoid operate switch permanently, clean switch.
Intermittent fault display F7, especially after rewind Check take up motor date code; less than 06/84 replace the motor.

VS300:

It can happen that the drum motor is going in reverse. Do not panic as this is symptomatic of a fault concerning lack of information to the microprocessor. Check the initialisation. Cause may have been capstan motor seizing and interface IC failure. Use the A5 program to operate without a cassette.

Hitachi

The basic design of Hitachi videorecorders is common to a wide range of models. The following symptoms and hints are applicable to models VT8000–VT9700 unless otherwise stated.

Cassette lamps A standard filament bulb is used in models VT5000, VT8000 and VT9000 series, whereas VT6500, VT7000, VT11, VT14 etc are fitted with an infra-red double LED emitter; this can be quickly checked with a TV camera, which is sensitive to IR light. IR remote control guns can be checked in the same way.

No luminance on record or playback Faulty filters CP205 and CP206; confirm by checking signal level on IC202 pin 14. If correct, check further downstream. Modified filters are available under part nos. 5162344 and 5162345.

Failure to record picture Failure of IC201, giving little or no FM carrier signal at its pin 25.

No visual search, IC901 pins 3 and 4 high Faulty syscon micro IC901.

No UHF output from RF convertor O/C choke in 9 V power rail within the screening can. Generally caused by the adjustment screwdriver shorting across the trimmer capacitor.

Failure of tuner/IF block Each section can be replaced separately. Tuner part no 2423461; IF part no. 70996666. Check screen connections in IF module for dry joints.

Audio buzz or instability during replay Replace relay RL401.

Squeaking and audio distortion Audio/sync assembly vibrating. Suspect ones bear no. '671'.

No take up of tape Faulty brake solenoid transistor Q56.

No play or intermittent stopping Capstan motor bearings stiff. Lubrication will temporarily cure the problem, but motor replacement is better.

Loading/unloading problems If RO81 is found o/c replace it by posistor PTMAR2R2M, part no 0249799. Also check IC 905, IC906, Q901, Q902, motors, IC901, afterloading switch and unloading switch.

VT5500:
No clock R954 o/c.

VT8000/8500:

Will not switch on to operate Check 'on' signal from microprocessor to power supply.
Stops when cue/review selected Check IC901.
Tracking errors and tape damage Check the impedance roller next to the erase head.
Failure to complete loading and operate the AL switch Change the drive belt.
Erratic systems control and nonsensical operation functions Usually due to the main microprocessor IC901.
No E-E signals, 'not P/B 12 V' supply missing R069 (system control panel) o/c.
Intermittent colour Replace IC203 and check Q225.

VT8300:

No functions and stop light flashing Cassette lamp failed.

VT8700:

Losing time Unplug from the mains for a while and then reconnect and recheck.
Tape damage after rewind No cassette lamp lit, check R653.

VT9300:

Poor take up and tape chewing Replace idler wheel.

VT9700:

Intermittent 'on' operation, operate light intermittently on Check 9 V regulator.
Intermittent slow speed Change capstan motor.

VT11:

Tape damage Change reel idler wheel.
Runs for a few seconds in rec/play then stops Change threading motor and belt.
Poor tracking with wide noise bands This symptom is often accompanied by intermittent stopping. Check for assymetrical drum flip-flop waveform due to failure of diode D481.
Random selection of functions Faulty diode D453.
Poor take-up or intermittent stopping Replace idler pulley.

VT14:

Enters record mode when switched to timer, IC904, pin 15 high   Failure of IC904.

VT19:

Intermittent clock display/reset   Faulty Q1795.
Failure to recognise SP and LP modes   Check IC908.
Audio motor-boating on replay   Change R426L and R426R to 680R. Add diode, cathode to junction R426L/R422L/IC405, anode to base of Q415L. Proceed identically for the RH channel. Transistors now turn full on in playback, shutting off the record amplifier. Diode anode is the marked end.

VT33:

Intermittent loss of colour playback   Check 9 V supply to Q217 and replace IC 203 (HA4239).
Failure to thread after the cassette is loaded   Check motor drive IC 902.

VT57:

Capstan motor intermittent in play   Two causes have been found, posistor PH1151 and IC1151.

VT63:

Noisy capstan   Check the capstan flywheel for wear and renew and/or grease.
Threads up and then stops   Check 12 V power rail to motor drive circuits.

VT64:

Cassette loading problems or the machine indicates a cassette in when none is present, or it will not re-accept a cassette once one has been removed   Replace both end sensor opto transistors, take care in re-alignment. Check also that the cassette housing has not been mis-aligned if a tape has been jammed.
Squeaking noise in FF or rew   Grease the capstan flywheel lower bearing and check the support bar for wear.

192

JVC, Ferguson

Most JVC videorecorders have Ferguson equivalents, and in each case JVC model numbers will be given, followed by the corresponding Ferguson ones. The first to be discussed will be the original 'mechanical' machines, recognisable by their finger-shaped operating keys.

HR3300/3330/3660 (3292/3V00/3V22); Akai 9300:

Certain models of JVC HR3300 (Ferguson 3292) used a different lower drum assembly to that fitted to later machines, and require the type of video head assembly having a PCB connection panel; JVC type PU31332-G10 heads will fit this range of machines, though stocks are limited. If it is required to fit the alternative head type (plastic-covered pin terminations) the lower drum unit must also be changed.

When later-version heads are fitted to early models the following modifications may be required: Record/replay panel – change R19 to 4K7, R54 to 2K2, C45 to 39pF. Set FM record level (TP1) to 2.2 V p–p. Y/C panel – change R33 to 470R.
Goes to stop when 'play' selected   Check cassette lamp.
Threads then unthreads when 'play' selected   To forestall operation of the stop solenoid during investigation, ground X5 base (mechacon panel at machine front) then check as follows: no flip-flop signal at mechacon panel due to faulty servo IC AN318 or missing head drum tacho pulses; take up spool or tape counter stalled, stopping magnetic rotation detector; defective 'start' and 'end' optical sensors; faulty after-loading (AL) switch; C12 (mechacon panel) low value.
Threading mechanism jammed in play   Replace capstan drive belts.
No sound erase, colour fluttering on new recordings   Faulty slide switches on the audio/sync panel.
Difficulty in setting servo discriminator amplifier after motor change   Incorrect instructions in service manual. For early capstan servos calling for a 33R resistor, and for drum servos calling for 100R and 5R6 in the setting-up process, use a 2 V supply, with machine unpowered. Adjust R76 in the capstan servo for ±zero current on meter 50 µA range; likewise adjust R111 in the drum discriminator. Connect as shown in the S.M.

Later models (HR3300 and 3V00) require no external components for setting up; an accurate digital voltmeter is required. Again use a 2 V external source (machine unpowered) for drum discriminator adjustment, setting R52 for a reading of -10 mV with the equipment connected as specified. Allow the drum to turn then stop it several times to confirm the reading. With the discriminator correctly set, carry out the sample position adjustments.

Model HR3660/3V16 has a braking transistor X7 for the drum motor. Remove it before setting discriminator pre-set R53 for -10 mV with 2 V applied, as above.

Head drum and/or capstan continue to run after unthreading On HR3660/3V16 this is commonly caused by failure of servo IC BA841. In other machines suspect gate failure within IC3.

Intermittent servo lock on record, no control track pulses recorded C207 (IF panel) low value, causing field sync pulse distortion.

Picture wobbles laterally, long drum servo settle-down time Ripple voltage across drum motor excessive (it should not exceed 300 mV) due to faulty drum motor. Similar symptoms arise from an under-damped servo due to low value C33.

Repeated failure of 3.15A fuse when play selected Collector-emitter short in transistor X1 (drum servo). R80 may also be damaged due to excessive current, leading to unstable servo lock. If X2 fails check drivers X10 and X12 to prevent 'run on' after unthreading.

Severe picture noise at top Insufficient back-tension (tension arm spring disconnected?), dirty or maladjusted entry guide. On one occasion the entry guide could not be set due to the head drum assembly being tilted – a foreign body was lodged beneath!

Periodic misregistration between colour and luminance image Replay of a self-recorded test card shows lateral colour shift at intervals of minutes, but tape replays satisfactorily in another machine. Tape tension variations are responsible; check back tension (28–45 gm/cm) and take-up tension (80–220 gm/cm). Note: the 'Tentelometer' guage is calibrated in grammes. To convert to grammes divide the specified 'gm cm' figure by 3, where 3 cm is the extra radius introduced by a full tape spool.

White smearing and odd black lines on replay; known good recordings replay well Worn video heads may be responsible, but this has been traced to a faulty modulator unit giving an assymetrical and distorted FM carrier.

Poor streaky picture with black speckling Results similar to those from a worn head, but due to failure of IC9 (luminance FM panel).

Low contrast, foggy picture Faulty IC10.

Herringbone patterning on colour during replay of own tapes Maladjustment of chroma record level control. Maladjustment of **replay** chroma level control (pre-amp panel) will cause fine symmetrical patterning on all replay.

Replay of self-recorded tapes poor and grainy; recordings play OK elsewhere, and pre-recorded tapes reproduced well This is a difficult one to come to terms with! The trouble generally lies in components used in both record and replay, i.e. LPF2 (PU31932-2), EQ2 (PU31933) and the head drum, where worn head chips cause low tape penetration; a healthy zizz should be audible as they scan the tape.

No colour on replay There are many possibilities here; proceed as follows: Check TP215 for 1.1 V p–p; if low replace crystal 202 (4.433619 MHz). Monitor replay colour path via 2-line delay comb filter; X204 has been known to fail. Check TP222 for drum flip-flop signal; if absent check connector 81–83. No 2.5 MHz output on pin 17 of IC208 – change IC. No 625 KHz on pin 4 of IC207 – change IC. No 5.06 MHz signal at TP219, use double-beam 'scope to check phase lock between signals on TP218 and pin 2 of IC207. If they are synchronous the AFC loop is locked; suspect BPF203 or the 4.435571 MHz oscillator. An accurate frequency counter will be required to check the latter. Little or no signal on IC202 pin 11, same on record – ACC failure due to a faulty IC200 (AN305).

No colour Check IC202, AN305.

Snowy replay, heads and FM waveforms OK Check IC7, SN76670.

Fuses blowing and o/c 4R7 in series with capstan motor Replace motor.

HR7200/7300/7350/3V29/3V31

The JVC HR7200 is a basic machine with a single event timer and mains 50 Hz-synchronised clock, whereas the HR7300 has a multiple-event timer and a crystal clock reference. Model HR7350 is similar to the 7300 but with stereo sound. Model HR7200 underwent several changes during its production life. The motor drive amplifier panel was originally mounted between the cassette compartment and the front panel, and started out with discrete transistors, being subsequently modified to use ICs. Later this panel was dispensed with altogether, and the fully-integrated components relegated to a panel mounted on the left side of the recorder.

The head drum assembly was modified to improve the tape lead-in path; this affected the drum heater, resulting in two drum types – be sure to order the correct one. To improve screening from adjacent TV sets sheilding cans were fitted over the audio/sync head and video head wiring connections.

Major changes were made to the luminance signal circuits to improve visual noise performance, involving a new noise-cancelling subpanel mounted beneath the tuning panel – see 'Recording and Replay Advances', chapter 3.

Cassette lamp failure. Caused by plugging into the mains with the operate switch 'on'.

Erroneous record selection from wired remote control Add 0.1 μF capacitor across the 'channel' key contacts.

Reverse visual search to fast Motor control buffer Q16 leaky. R48 (10R fusible) in this part of the circuit has been replaced in production by an integrated 'circuit protection device'.

Drum motor does not run, drive IC overheats One Hall-Effect switch low output, necessitating replacement of motor complete.

Poor drum servo lock Check pulse position pot R207 (drum discriminator circuit) then suspect excessive current drain in IC HA13008, betrayed by high ripple voltage on its supply pin.

No drum pick up head pulses at pin 16 of IC3 Change HA11711 and check invertor Q3.

Drum motor runs after eject Hardened grease preventing release of UL switch by slider below left side of cassette compartment. Re-lubricate.

Erratic capstan speed changes Check for presence of 126 Hz FG waveform and correct IC1 operation; also check replay CTL pulse level. On one occasion, C001 (across motor) was found leaky. A faulty capstan motor can give rise to warble on sound.

Capstan motor does not run Early machines using front-mounted MDA panel: dry joints on Q206 etc. at MDA PC board.

No fast-forward or rewind; failure after initial try D17 o/c.

Enters 'pause' when 'record' selected Syscon micro IC1 (uPD553C-164) faulty.

Extreme wow and flutter, capstan motor OK C239 o/c, deleting AGC control to overload the modulator; high E-E signal level.

Intermittent operation of pinch roller solenoid Hardened grease impeding slider along

right hand side of deck.

Poor tuning Resistance drift on presets due to the whole assembly moving with heat; s/c D8 preventing conduction in Q8; s/c or leaky Q8, leading to damage in regulator IC3 and excessive (45 V) tuning line voltage.

Erratic lines on playback picture C55 (pulse sharpener) faulty, deleting capstan sample at TP5, and CTL pulses at Q1.

Replay picture gets progressively noisy Q102 leaky, causing high E-E level and gradual decay of playback FM signal.

No picture record Faulty D208 turning on Q209 to clamp FM signal to ground, indicated by no signal at HA11724 pin 1.

Poor colour, loss of colour sync No 4.43 MHz signal at TP403 due to o/c C468.

No colour record, no apparent fault in chroma circuits D401 cathode voltage down from 9 V to 2.9 V due to faulty colour/mono switch.

Poor colour replay of self-recorded tapes, OK on others No burst gate pulses at pin 3 of IC401 due to o/c L407. This has the effect of raising signal level at IC401 pin 16 (ACC stage non-operational)

No sound Audio muting point (IF IC pin 6) high due to o/c coil T4.

Sound buzz on E–E O/c C47 in muting circuit permitting video sync pulses to reach pins 4 and 6 of IC2.

No clock display No clock pulses on driver IC due to leaky C45; in some machines it is designated C407.

Incorrect display, time setting impossible IC uPD553C-100 faulty.

Clock losing time, crystal oscillator within 3 Hz of 4.194304 MHz SM5502 output at 1.2 sec instead of 1 sec – faulty counter IC.

No timer recording, threads up to 'record pause' but no more TMS1024 data selector IC faulty, micro IC ok.

Fast drum rotation Failure of IC201, VC1029.

No audio record Zener D7 s/c.

Drum speed not correct Check IC13.

No FF/rew or visual search Check CP2.

No clock display Check power rails particularly D13 and Q5.

Low drive voltage to the capstan motor May be due to the speed control IC1, a VC1029.

No E/E picture Check Q221.

Intermittent power supply Check Q101.

Enter pause when record selected Main control micro may be faulty.

Threads up and unthreads again Most probable

cause the threading motor belt. Check for system control problems such as no flip-flop due to o/c drum pick up head.

Threads up and runs for a few seconds before unthreading again when hot Check loading motor belt for slipping and turntable optical couplers.

Static spots on replay Could be due to the drum flywheel chaffing on cables below it.

Channel selection erratic Check IC201 and D221.

Channel 1 only selected Fuse F5 blown, check Q8 and IC3.

Very high E/E signal Can be due to the playback 9 V being present in E/E mode, if so check Q107.

Critical tuning and smeary video Change the IF IC AN5111.

No E/E sound Check D14.

Intermittent stopping Can be dry joints around Q216 or Q17.

HR7700/3V23:

Power supply Fuse F5 blown by s/c C36. Fusible resistor R38 was replaced in production by a wire link; if this is done subsequently C75 on Servo I PCB must be removed. This can create servo problems in slow motion.

Rapid cassette load then eject X1 s/c on PSU panel.

Poor eject of some cassettes Roughen up the surface of the rollers above the cassette housing frame with emery paper – or replace the roller assembly. If there is any sign of creasing in the tephrone strip sliders (on the bottom of the cassette carriage) replace them; part no. PU49001A.

Pinch roller does not pull in Transistor X5 o/c.

Tape fails to thread X7 o/c in loading motor circuit. Carefully check also X1, X15, X21 and X47.

Difficulty in diagnosing syscon faults In order to maintain operation of the mechacon under fault conditions, the 'interrupt' signal at IC1 pin 6 can be inhibited by grounding pin 9 of IC17; only cassette lamp failure can now invoke 'interrupt'.

No fast forward or rewind in 'cue search' mode only – otherwise OK X49 s/c, holding pin 26 of IC12 (data selector) permanently low.

No reel motor functions, X23/24 and 6 associated transistors dead Usually X2 and X46 survive; check and replace all transistors in 'bridge' motor drive circuit, replacing as necessary. Overheating

of X18, with X46 hard on and IC1 ok, was traced to IC9 in the logic gating circuit.

Goes to stop 3 seconds after selecting 'play', motors not running Check motor control voltages on pins 38 and 39 of IC14, servo 1. Replace IC14 if control outputs do not respond to control input at pin 2.

Poor take up and intermittent stopping Suspect reel motor. Check TP3 on servo 2; if waveform incorrect suspect X12 and the condition and alignment of the pick-up head. If all is well there, check TP5; an incorrect reading here should lead to a check of X5 then IC3.

If IC17 pin 8 changes state, it is likely that one of the spool rotation detector outputs is going high ('no rotation' indication) or that the drum flip-flop signal is missing.

Intermittent shutdown, no power 'on', front panel controls not working Random 'on' and 'off' commands can stem from a faulty X1 transistor on the key scan panel. Failure of front panel keys (remote control working) should lead to a check of data output from IC1; if absent X1 is probably o/c. If it goes s/c, the front key panel takes priority, inhibiting remote operation (continuous data present) but manual 'power on' will work.

Incorrect operation of front panel control keys Check data selector (ICs 23/24/25/26. Check the clock signal on pins 1 and 42 of the micro; if present, but pulses are absent from pins 30 and 31 replace the micro.

No response or incorrect operation of the remote control, possibly upsetting front control key operation Check gate switches TA1 and TA2, within the TA57 IC; TA2 can also affect the 'tape remaining' display to flash segments erratically.

No still-picture or slow-motion, correct LED indication, sound muted but normal play pictures displayed Check IC14 (BA841) and its CTL pulse feed, also X41. If noise bars appear on the picture check the stepping pulses to the capstan motor.

Drum motor running fast Drum FG decoupler C52 leaky.

Both servos malfunction Check 4.433 MHz crystal oscillator on servo 2; measure TP2 waveform for 8 ms low and 12 ms high.

Intermittent loss of tracking with sound pitch variation Check IC1 on servo 1 for correct pulse level and absence of spurious noise.

Vertical ripple on stationary pictures Regulator X3 s/c, raising drum servo 12 V supply line to 22 V. Low 12 V line may be caused by a faulty zener D13.

Negative picture and buzz on E–E operation Modulator overload due to low supply voltage – X12 faulty.

No E–E picture through X3 on junction PCB IC14 (Servo I) faulty.

No UHF output Check X12 before condemning the expensive RF convertor (modulator).

Erratic channel storage Carefully check all joints in relevant area of memory PCB, then if necessary replace MN1208 IC. A modification may be required: Add a 6.3 V zener diode, pointed end towards R63.

One of clock or timer display missing Change CF1 and CF2 from red type to new blue type ceramic filters; if the problem remains, or erratic operation ensues, change C1–120 pF, C2–470 pF, C3–120 pF, C4–22 pF and R13–560 K.

Clock/timer display panel fails to illuminate Check DC/DC convertor negative output voltages. Check 10 V power rail and 'power on' reset on the T/T panel (i.e. IC10), and 'power off' input from D4 on PSU panel.

No clock count Check and adjust crystal oscillator IC3 on T/T panel.

No E/E sound Check for presence of audio at the input of IC1 on the audio PCB, check status of X4 for rec/play switching, also check and replace the audio input switching IC on the junction panel.

No still frame or slow motion Check the AN360 IC.

Weak replay picture with picture jumping Check C44.

HR7650/3V31:

Failure to store selected channel MN1218A memory IC faulty.

No video record FM carrier absent at TP222 and TP223 but present at pin 1 of IC205; Q235 (in record output stage) s/c.

Intermittent blank or flashing TV screen in E–E Defective C19 (at pin 23 of IC1 in tuner/IF block).

Tape speed varying This can be the servo IC6, AN6341 or the switching chip IC7, a TA4066.

No fast forward or rewind Check IC5 and Q18.

Intermittent audio erase Replace oscillator module and switching relays.

Broken up picture on replay Check Q103.

Intermittent erase and audio record Replace bias oscillator module with later type and both switching relays.

HR7655/3V32:

Distorted sound in E–E, volume varying Re-tune T5, check Q6 operating conditions.

Intermittent colour R339 colour AFC off adjustment, set for 625 kHz.

HR2200/3V24:

Alarm indication, front panel lights strobing Check all leaf switches and voltages on the microprocessor, static can cause the microprocessor to fail by blowing an input, check that data inputs are 10 V when open to ground.

A pale picture and lack of AGC Can be due to X5 failing to clamp to a DC level.

Tape edge damage Can be due to a bent pinch roller.

Intermittent power supply off the tuner or power unit Check DIN plug and cable connections.

HRD110/3V38:

Tuner button lights stay on when the machine is switched off Check Q5.

Cassette loading compartment damaged Replace the whole unit.

Operational problems with the end sensors Check the de-coupling capacitors for s/c first, also applies to housing down detect switches.

Intermittent play or speed variations Check After Loading switch.

No colour Check 5.12 MHz and transformer T1010.

Intermittent failure to respond to the front buttons Check data pin on rear DIN socket for shorting to ground.

HRD120/3V35:

Will not power up Check transistor circuit protectors or power supply switching relay mounted on power PCB.

Threads up and then stops when play is selected Check AL and UL switches.

No fast forward or rewind Check signals around IC206.

Tapes not being wound back into cassette, sound warble, noisy motor Replace capstan motor in all cases.

No functions Check 9 V supply rail and circuit protector CP2.

No colour Check crystal XB401.

No colour playback   Check the HA11741 voltages and waveforms, C433 can be s/c or BPF401 faulty.
Drum servo hunting   Check the value of R446.

HRD140/3V44:

Intermittent erase and audio record   Add 5.6 nf capacitor across C23.
No 'on' operation or drum rotation   Check CP4. Check also that the power 'on' signal operates on pin 1 of the control microprocessor: if there are no voltage changes replace the microprocessor.
Takes in cassette but no further functions   May be due to D628 being leaky and holding logic level incorrect.

HRD565:

No hi-fi muting in visual search   May be due to glue on pin 9 of IC 208 preventing normal operation.

HRD725:

No clock display, or all segments lit   Check 12 V power rail at Q2. If this is high (18–22 V) the following must be changed: Q1, Q2 and D2 in the power supply. Then check *and* replace on the tuner/timer PCB Q3, D13, D17, D18, D19, and D20. C13, C14, L1 and also IC1. Do not forget to disconnect the back-up battery during component changing!
Intermittent tape loading or no reverse cue   Re-grease loading gear wheels and/or replace loading motor assembly.

GRC1/2 3V41/50:

Alarm modes

In the event of an alarm mode being entered a flashing indication is given by the function LEDs. The problem area is defined by the function LEDs which accompany STOP. In some instances the Videomovie will run, although it may be only for a short time. The alarm indication will persist for approximately 5 minutes, after which time the overtime circuit resets the power relay and power is discontinued.

Indication	Problem
STOP	The microprocessor is not receiving the correct mode control information from the mode control cam switch. Either the mode motor is not in the correct STOP home position or the cam switch is faulty.
STOP/REWIND	The supply reel is not turning.
STOP/FAST FORWARD	The take up reel is not turning.
STOP/RECORD	Capstan flywheel FG signal is not detected.
STOP/PLAY	Mode motor may not be running in the loading direction, or no output has been detected from the cam switch.
STOP/REWIND/PLAY	Mode motor is not detected running in the unloading direction.
STOP/PAUSE	Drum flip-flop signal not detected, drum may not be turning.

GRC1/2:

Tape damage with no tape take up   Replace take up idler arm. In the GRC1 the take up idler arm can easily be removed, however, in the GRC2 (and later versions of the GRC1) a large washer is fitted beneath the deck which makes removal difficult, it is better to replace the rubber tyre only.
Cassette lid open or permanent eject mode   Unplug mode motor and reset to home stop position with an external supply. Check CP1 and then other power rails including the 5 V rail on the servo section TP301. If no reason can be found for its failure replace the mode cam switch. This fault symptom can occur if the threading gears are out of timing.
No drum servo lock in record   Check drum servo waveform, if the sample pulse is missing check for low level syncs on pin 9, check C3 in sync filter circuit.

No replay pictures Check IC12 in the FM demodulator circuit.

No playback audio Check voltage on pin 19; if less than 1.8 V change C31. Check voltage on pin 13, if it drops significantly between E/E and playback then change C27.

No E/E audio Check voltage on pins 12, 13, 14: if less than 0, 1.9, 1 V (respectively), check bias voltage on pin 16 and pin 20, if pin 20 is <4.3 V change C15.

No power up and no indications Check power input and CPs. Check syscon microprocessor reset circuit, C33 can hold microprocessor in permanent reset.

Power will not stay on, with a cassette inside If the STOP/REW/PLAY alarm is initiated then the tape is jammed and the mode motor is stalled.

No E/E monitoring picture Check camera output and AGC output (pin 5, IC5). If the camera output is present but the AGC is not, change C25 and then recheck, if the result is still no output change IC5.

Supply or take up guide bent is displaced Replace all parts especially slider springs and check for alignment.

Viewfinder monitor not working Check brightness and contrast control settings, check and replace high voltage unit.

Intermittent white noise on picture or with 12 Hz flicker Replace video heads and/or check preamp switching pulses and connections within screening can.

Green screen This represents no output from the camera. Failure of pre-amp IC1 or failure of the tube HV unit.

Colour unstable in replayed copies with drop out to monochrome AFC adjustment very fine and colour lock just hanging on: change IC 8 MCM1068A (very expensive IC).

Colour flicker (at about 12.5 Hz) on own recordings Colour signal low on one head record to replay, if it does not respond to cleaning, replace video heads.

GRC1:

Intermittent capstan servo Check C17.

No E/E pictures Check AGC output on IC5, MC5575.

Excessive symmetrical patterning in replay IC2, MCM 1048B.

No zoom, changes focus instead Lens set on Macro.

Sound warble Replace capstan motor, check belt also.

No capstan servo in replay Check CTL pulse level. Clean CTL head. Check C115 if IC104 is turned off or voltages wrong.

Fast play speed uncontrolled Check capstan motor FG signal for level and change C109.

Switches off immediately after switching on Check all power supplies and circuit protectors. The mechon PCB can be tested on its own, out of circuit, with the control panel and substitute power supply, and with PCB test units connected. Current consumption is 1–2 mA. It can be switched ON by the test PCB but not OFF. The power relay will reset when the power connection is disrupted, if this does not happen the power off circuit up to Q22 can be checked.

No playback pictures, blank screen No PB supply line as IC10, MC5365, pin 6 o/c, if a replacement also goes o/c, then check IC2, MCM1048B for excessive loading.

No record function or record/pause (assembly edit) Tape can go into assembly edit but continues in reverse mode until STOP/RECORD alarm. This indicates no capstan flywheel FG signal, check C24 by replacement.

No reverse functions review or rewind Check and replace Q109, Q110, Q113 and Q114.

GRC2:

Intermittent capstan servo lock Modify capstan servo according to data sheet to upgrade against tolerances.

Full white screen Check fader circuit resistors for dry joint.

Intermittent yellow tint Check additional components connected to IC4 pin 20 on the camera video process PCB.

Autofocus not working Check lens is not set to Macro. Check for CB, BF and FF signals on IC5 on the EE/IND PCB. If one is not present change autofocus assembly and set AF back focus with special tool.

'Battery' indication permanent display in the VF monitor Most probably due to trapped wiring, remove camera unit, redress cables and recheck. No VF function indications, check VF socket PCB and redress camera connections as above.

Capstan motor running slow, accompanied by shut down Check CP3 for failure. If there is no apparent cause, remove camera unit and check/redress cabling.

Mitsubishi

HS303:

Intermittent clock Tantalum capacitor C8F2 leaky, causing 'INIT' pin to rise from 4.7 V to 10 V.

Intermittent audio hum Replace head switching relay, K3F0.

Intermittent no audio record Check wires on plug VK.

Tuning drift and whistling on sound Check tuning voltage for interference.

Cassette tray loading problems Check the white nylon carrier bearings.

No replay colour C6D1 s/c.

Intermittent black strakes on peak whites Check IC2B0, white clip control.

Unstable drum servo Can be due to C4B8 low in value, this can only be proved by substitution.

HS304:

Intermittent 'thread' action when 'eject' selected Check and adjust AL and UL microswitches hidden below the deck in the far left hand top corner.

Will not play prerecorded tapes Carefully reset tape guides and seal with nail varnish.

Failure to return to normal play speed after cue or review Check hall effect inputs to IC4A0, if they are different replace the motor.

No capstan motor rotation Check IC4A2.

HS306:

No cassette loading but will eject Check capstan motor drive IC, STK6962.

No colour record Check IC6A0.

Intermittent mains fuse blowing Uprate to 630 mA.

HS310:

Intermittent tape transport in timer or normal play Faulty cassette lamp.

Head drum running at wrong speed Faulty reference crystal X6F2.

No colour in replay Check PCB solder joints around crystal X6A1; if necessary check frequency setting of subconvertor VC6A1 and VC6AO crystal trimming capacitors.

HS318:

Changes in booster amplifier gain when the VCR is switched on Can be due to the 9 V regulator being faulty.

HS700:

Drum not turning and drive IC HA11715 hot Check IC4P0.

Machine runs for a few seconds then stops Drum flip-flop may be missing, check the wiring to the rear mounted still frame adjuster.

Murphy – see Sanyo

Panasonic

NV333:

Will not tune in to stations Check tuner.

Intermittent audio erase and record Short circuit R4049.

Intermittent chroma record Connect a fast acting diode (MA161 or MA162) between pins 11 and 12 of IC6003, cathode to pin 11.

No clock or capstan motor Check 18 V supply and Q1002.

Intermittent record or play Check the loading belt for slipping.

Poor sync signal on replay Check IC3003.

Poor capstan lock on playback Check 9 V supply and replace Q6003 if faulty.

No audio record although erase was present Check D4001 and Q4003. This can also be due to the power supply being higher than 9 V.

No colour record D8005 leaky bringing on the colour killer.

A case of permanent eject Check Q6028.

NV366:

Incorrect capstan speed Check TP2007 for FG pulses and motor bearing.

No clock display Check the 36 V supply rail or change the clock IC.

Intermittent poor replay after selection of cue or review Replace both relays RL3501 and RL3502.

NV370:

Poor rewind or low take up tension Fit modified version of idler pulley.

Noisy sound recording Check IC4001 is switching properly.

No display Check power supply −30 V and Q1102, D1109 and R1102. Also check Q6501 and I6501 by replacement.

NV777:

No line lock Check TP2006 for PG pulse; if absent, replace the drum.

Intermittent cassette motor or loading motor Replace IC6004 or IC6005.

NV7000:

No audio E/E Check for incorrect audio muting and IC6010.

Clock display on switch on consisting of a top row of dots and a lower row of figures IC7505 is held in reset, check zener diode D7540.

No functions Replace cassette lamp.

No play after selection of cue or review Replace loading belt.

No play, no take up drive Replace play belt (below cassette compartment).

NV7200:

Timer not programmable, Clock display erratic Change clock IC.

Philips

VR6520:

No capstan drive R96 to pin 24 of the AN3822 or the IC itself.

Cassette tray or threading motor drive absent Check R1101 for o/c.

VR6660:

Faulty clock display Check IC2, SN75518N.

VR6460:

White spots on the screen May be due to the drum motor creating static.

VR2020:

No pinch roller after threading up Check T7007 and T7008.

Pinch roller goes in and then drops back If pin 15 of U220 is lower than 5 V, change T7002.

No E/E sound Change IC7051 in U160.

Unstable drum servo operation Check relays 1001 and 1002.

VR6462:

No functions Check supply rail for 17 V, if higher check D6103.

No rewind, fast forward or intermittent play with tape damage Brake solenoid sticking so replace it. As a cross check listen for a defined 'click' when changing from FF to rew or vice versa.

Clock failure with erratic display Replace clock IC.

Sanyo

VTC9300:

Machine dead Replace 12 V regulator transistor with TIP41 type; ensure that mechanical mounting is good and that good thermal contact from transistor to heat sink is achieved. Sometimes the transistor will go intermittent o/c. More rarely it can go s/c, risking damage to its driver transistor and some ICs fed from the 12 V rail, notably the audio detector IC on the tuner board.

Machine stops within seconds of selecting 'play' Pinch roller not closing due to insufficient holding current — failure of Q810.

Tape counter counting UP during rewind Incorrect 'down' command to TMS1070 micro due to leaky C1618 upsetting Q1602 function.

No pause function Q810 o/c.

Excessive picture wobble on playback Faulty C134 (servo damping capacitor).

Incorrect tape speed, excessive braking coil current An example encountered proved difficult to diagnose; timing pulses at TP 104 were 40 ms in record and 41–42 ms in playback. During playback the braking coil current was in excess of the normal 50 mA; the capstan was running slow due to stiff bearings. Lubrication restored normal operation.

Grain on 'off-air' pictures routed via VCR at UHF This model has no aerial booster amplifier, and introduces about 3 dB loss in the TV aerial lead; in fringe reception areas this causes noticeable deterioration.

Clock gaining time IC counting mains spikes. Add suppressors in the form of two 0.01 μF capacitors to ground – one from Q1608 base, one from pin 3 of Q1618.

Timer set fault A machine was encountered in which turning the power switch to timer position (timer having been set) had the effect of leaving the machine on with eject solenoid clicking. Replacement of timer panel had no effect on the fault symptom, but the operating voltage for the timer IC was found to be 0.4 V high at 9.8 V. The fault was due to a leaky 15 V regulator transistor raising the 15 V line to 20 V; this overvoltage had no other apparent effect.

Intermittent recordings or playback Pinch roller solenoid not operating or dropping out. Check Q810 and Q816 also replace diode D814.

No servo lock in record, playback ok C108 faulty.

Unstable servo in playback only Check capstan flywheel bearing and belts, if all else fails de-gauss the CTL head.

No picture in record although E/E ok Check Q4 on W1.

VTC5000:

Reluctance to loading May be due to wear on the worm gear clutch.

Will not playback own recordings but others ok Check the CTL replay level and C4505 on the sub PCB (if fitted).

Clock jumps time or resets to zero after a timer recording Replace C3308.

No tape take up Check reel motor for dead spot or IC3006.

Blown mains fuse Check regulator IC5101.

Poor capstan servo lock when replaying own recordings Add 390R across R4515.

Tape looping during unthreading Replace reel drive components and change R3049 to 1R0.

Intermittent stopping in record or playback Faulty reel motor if take up reel stops. If not check output of reel sensor optocoupler for 6 V p–p; if markedly lower clean and if necessary replace optocoupler.

Random mains fuse blowing Replace mains filter capacitor by RS Components type 114–659, 0.0047 μF. Fit correct anti-surge mains fuse.

Random diagonal lines of white spots on replay Oxidisation of earthing contact on FM pre-amp screening can. Fit a solder eyelet tag to the screw retaining its support bracket and solder a short grounding lead between the tag and the screening can.

VTC5150:

No colour in playback Check signals and DC voltages on pins 26 and 27 of IC11009, suspect R1297 if they are incorrect.

Faint vertical lines on replay Move the position of grey lead JW18 between the two delay lines.

5300:

Intermittent stopping Check tape counter drive is free and change belt.

VTC5400/Murphy MVR7007:

Many of the above comments for VTC5000 apply to this and other 5000-series models.

Intermittent threading or unthreading Two factors responsible, sharp corners on cam profiles around the threading ring (file smooth and lubricate) and a slack loading belt. To replace the latter, proceed as follows: remove cassette compartment, take-up turntable, large black plastic shield and one drive pinion; a tension spring below the deck will fall off at this point – make sure it is reconnected during reassembly.

It can happen that the loading motor develops a 'dead-spot' to give the same symptom, but this situation is betrayed by an absence of motor noise in the fault condition.

Capstan speed variations Check C4040.

Sharp

General: The fault reports given here are for individual models, but electrical and mechanical similarity in many cases means that those given for one model can apply to others.

Note that the cassette lamps for Sharp VCRs are 5 V, 100 mA not 12 V, 60 mA as usually found.

VC381:

Take-up reel stationary in record, OK in playback VS-H line high in playback whereas it should only be so in visual search modes. IC7754 faulty, with pin 1 stuck high.

No clock reset Change MP2812.

Problems on the Y/C PCB from cold Check C438 and C439.

No colour record Check 5.06 MHz filter, FL503.

Poor take up and rewind Replace idler pully and/or reel motor.

Intermittent sound recording   Change relays and change C648 to 0.01µf and R693 to 10R ¼W.
No field lock in E/E, no syncs at IC402   Check C438.

VC386:

Head drum rotates too fast   Check for drum FG and C710 for s/c.

VC387:

No functions   Check the supply line to the microprocessor and then look for circuit protectors not shown on the diagrams.

VC388:

No functions and failure of regulator Q5002   Operate carefully; if Q5002 heats up then replace the microprocessor I5002.

VC390:

No erase   Check Q602 and Q603.

VC482:

No pictures when play selected   Traced to the record 9 V being present due to failure of IC802 which drives Q803.

VC581:

Erratic system control operation   Remove 47µf capacitor from across the loading motor before proceeding further.
Capstan rotates backwards   Due to a short between D7018 and link J20 to link J25.

VC2300:

Tape stuck in loaded position   Check Q914.

VC7300:

Loading motor continues to run in play mode   Main solenoid not releasing at stop mode – replace or lubricate plunger.
Motors pulsing   Check regulator transistor.

VC7700:

Weak or poor tape loading   Check loading motor and Q8091.
Intermittent recording   Check 1301, monitored at pin 8.
Threads up and then unthreads   Check slack sensor for error signal.

VC8300/VC9500:

Warble on sound   Dirty commutator in capstan and/or reel motor imparting 'lumpy' drive to tape. Replace either or both as necessary.

VC8300:

Threads up but fails to play   Check cassette lamp then if necessary remove tape deck to check connectors below solenoids.
Replay picture deteriorates at each playback   Partial erasure of tapes due to faulty control IC601 permitting erase oscillator to run.
Permanent operate light   Due to Q902 s/c.
Intermittent playback picture   Check connector CD on PWB-C.
Low motor power   Check I805; if it has failed check also the motor changeover switch before replacing the IC, the switch can fail and s/c.

VC8581:

Intermittent recording FM carrier   Check IC401.

VC9100:

Threads up and then stops after a few seconds   Check position of AL/VS/UL slide switch.

VC9300:

Intermittent stopping   Check and clean or replace reel drive idler; check reel motor for dead spot.
No rewind, tendency to stay in fast forward   Relay R775 sticking.
No rewind or no fast forward, sometimes fuse blows   Change relay R7751.
Cassette detector switch breaking on the top of the cassette housing   Prone to this fault – replace.
No take up drive with FF, rew and slow   Q7754 o/c.
Variation in capstan motor speed   Check R745.

No FF, rewind, or visual search Replace idler wheel first and if problems persist then replace motor.

Intermittent stopping Can be caused by the tape counter as the take up rotational detector is driven by it. Check belts and the worm gear.

Intermittent fuse blowing If no reasonable reason can be found check Q903 and D901 by replacement.

VC9700:

Threads up but fails to play Head drum not running due to L701 (13 V supply) o/c.

Incorrect tape speed, erratic servo operation Check for presence of 50 Hz reference, derived from 4.43 MHz crystal oscillator on chroma panel.

Poor line hold in search modes Off-tape control pulses varying in period due to faulty reel drive motor.

Intermittent sound recording Bias oscillator relays making poor contact — clean contacts.

Poor servo lock in playback Check 50 Hz lock on TP707 and Q710.

Insufficient take up torque Check drive wheels and then Q708.

Tuning drift with dim display Replace posistor PR6601 on audio PCB where the 30 V power supplies are generated; it is the same as used in Hitachi motor drive power supplies, being 4R7.

Sony

SLC5, SLC7:

Machine dead, mains fuses blown or R101 open Power supply failure. Use Sony kit A6738-159-A, replacing Q101, Q102, F006, C108, C110, C111 and R101, if C109 is not charged. Check IC1, D201, D202 and Q201. WARNING: C108 and C109 are charged to 300 V; discharge them via a resistor at each switch-off during diagnosis.

To work on the power unit out of the machine: connect mains supply via an isolating switch and variac to white and brown on plug CN0004; link red and white on plug CN0009 to put T4 into 240 V working; short together yellow and red on plug CN0015 to energise 'Ever 12 V' regulators. To power up, short together reference points 16 (blue wire) and 11 (red wire); this turns off Q201 by linking its base to the Ever 12 V supply. Always use Sony-supplied PSU replacement parts for safety and reliability.

Will not switch off into standby Q201 (SR08-2 board) s/c.

Poor rewind Fit replacement rewind kit A6706-348-A. If the fault persists the head disc may require replacement.

Squealing noise during thread and unthread Often associated with sound wow and flutter in replay, this is due to a faulty capstan motor; it is difficult to replace due to access problems.

Rewinds at threading completion Check end-sensor oscillator circuit. Alternatively, will not go into rewind — check start sensor circuit. The forward (end) sensor IC8 and rewind (start) sensor IC9 outputs are normally low, pulsing high when oscillation ceases.

Test signal present during replay D41 (YC-6 board) leaky.

No picture on playback IC1 (FM demodulator) faulty.

Intermittent muting on replay Reduced CTL replay level due to worn audio/CTL head. Increase replay gain by modification: On AS-3 panel change R123 to 2K7 and replace D30 by 180R resistor. It is better in the long term to replace the ACE head assembly.

Auto-tuning not working Faulty IC4 on CH-3 board.

All segments simultaneously lit on channel display Check IC7 (NAND gate) on CH-3 board.

Failure to select TV channels after playback Check IC1; if the channels are off tune it is likely that memory IC3 is 'soft' with age.

No colour in replay Check IC11 on YC6 board; this removes the pilot burst during playback.

No colour, ÷8 counter not working No 5 V supply due to failure of Q64 on YC-6 board. If 5 V is present check ICs, particularly IC9 in the counter chain on CB-1 board.

Intermitent loss of colour in record No signal out of Q55/Q56 due to o/c L29 in low pass filter.

Sound wow and flutter See under 'squealing' above.

No E–E audio, no audio record IC503 (tuner panel) faulty.

Intermittent selection of channels or channels selected being off tune and then on tune next selection Change memory IC, CX761.

Picture stretching to the right Servo unstable, check C11 for 20 ms period or replace.

Fuse F004 blown, clock display reading E:E, all timer leds lit, will not switch off Replace IC1 and Q3.

No E/E display except when play/pause selected Check power rails if playback/dub remains high then check Q423.
Eject and rewind but no play, or no rewind Check BX342 sensor ICs.
Channel indicating 18 (full segment display) Check timer IC.
Wear on the control track head A modification can be carried out; change R123 to 2K7 and replace D30 with 180R.
Intermittent poor recordings Can be due to a noisy TV/Camera switch.
No servo lock in record Traced to Q25 failing and allowing control track recording pulses back to the capstan servo.

SLC5:

Head drum reluctant to turn in play unless pushed D1 (AS-6 board) s/c preventing 'start' voltage of 10.8 V appearing at IC1 pin 18.

SLC6:

No eject Faulty 'cassette off' switch S7103.
Poor rewind Fit kit A6706-391-A.
No fast forward or cue functions Fast forward solenoid driver Q025 (DR-1 board) leaky. It is a special Darlington type.
No capstan drive Failure of 12 V regulator IC001; this increases servo IC2 (CX143A) supply voltage to 15V.
No E/E Check power rail supplied by Q01.
No reel functions Check R061 for o/c.
No capstan motor Check the power supply voltage to spec, error signal from the servo and Q022.
Intermittent operation of the threading drive, particularly when cold Replace all the belts and relay pulleys, also the phosphor bronze bearing between the lower reel motor and upper chassis, Part No: 3-671-122-01, not given in manual.

SLC6 MK11:

Tape damage at insertion of cassette, top half of old recording not erased during subsequent record Modify as follows: (1) Cut SS-9 board PC track between pin 23 of IC502 and R530; bridge the cut with a IS1555 diode, anode to pin 23. (2) Fit a IS1555 diode, anode to IC pin 30, cathode to R530. (3) Add a IS1555 diode, anode to IC pin 37, cathode to IC pin 2. (4) Fit a 3.9 K resistor across

R579. Note: it is sometimes better not to fit this resistor.

This modification works well, and also ensures that the tension arm is properly positioned during cassette ejection. This eliminates the 'half picture record' symptom arising from failure of the tension arm to route the tape correctly over the full-erase head.

SLC9:

Broken guide pins on threading ring Use Sony kit A-675-910-7A. A pair of end cutters will be required.
No clock or programme number indication DC/DC convertor (feeds fluorescent display panel) faulty.
Poor E/E audio Check IC520.
Cassette compartment loading gear drive fails Fit modification kit, Part No: A-6751-212-B.

SLC30:

Beat frequency seen on screen which is not due to co-channelling Check C319 in the power supply.
Poor rewind Check that the brake solenoid is working; it has an internal thermal fuse that fails.

Toshiba

V31:

AGC overload on modulator in E/E Probably due to PB12V present, check Q663.

V65:

No clock display Check Q401.
No power on Can be due to the main microprocessor, check for output from the microprocessor to the power supply switching between about 3 V and ground.
Intermittent erase Check contacts to erase head.

V71:

Erratic function operation Replace the mechon control cam switch.
Reel motor IC fails Ensure that the reel motor mounting plate is earthed by adding an extra screw.

V5470/Bush BV9600:

Poor rewind Replace idler pulley wheel and upper cylinder; video heads may require replacement.

Intermittent unloading or eject; unloading sluggish Replace loading motor assembly.

Drum motor continues to run after eject Check 12 V supply rails for power leakage. Check D998 (TV/VCR switch panel) for s/c.

Sound warble First replace take-up clutch. If fault persists suspect faulty capstan motor. Note that there is a selection of capstan motor pulleys; to avoid poor servo lock select one which gives correct free-running speed.

No picture record, Q404/5 not working Incorrect transistor bias due to leaky C419.

Tuning drift Faulty IC. May be ICA01 (TC9002AP), ICC01 (TMM841P) or ICC02 (TA7619AP).

No colour on replay Faulty pilot burst cleaner IC203.

No clock illumination Replace both fusible lamps, not an easy job!

Record button releases in timer recordings Ensure R619 is 330 K type; early models were fitted with 150 K here.

V8600, V8700:

Eject solenoid not working, no 'pull-in' pulse drive Q630 faulty, C620 o/c. Similar failures in corresponding circuits on the solenoid drive PCB (in front of cassette compartment) can upset operation of other solenoids.

Intermittent failure to record in timer mode Fit additional $10\,\mu$F capacitor across slack sensor delaying capacitor C628. Check for sluggish take-up due to oil-contaminated take-up belt. This can be checked as follows: put the machine into standby then 'power-up' whilst holding down the play button. If the slack sensor now operates check the timing delay capacitor in the slack sensor circuit also.

Operating keys not working Check input data lines (1–8) to IC601. ICL02 (front key panel) could be faulty.

Reverse functions do not work, or forward functions do not work Check start- and end-sensor levels; should be 3 V p–p at TP601 and TP602. If not suspect IC605.

PQS (Quick programme finder) insensitive Change C613 to O.47 μF.

Negative picture Q661 faulty.

No E–E pictures, no 12 V power rail Check R839 (22R fusible).

No channel selection ICA01 tuning control IC faulty.

Tuning drift Crack in PC panel print (12 V supply line) close to the tuner. Leaky zener D815 upsetting operation of tuning voltage stabiliser.

No colour, zero output at pin 15 of IC207 L215 s/c turns.

No colour, zero output at pin 6 of IC204 Check IC203, Q228, C246.

Intermittent no colour Check AFC setting (R258) then suspect IC208.

No colour in still-frame, cue and review If output on pin 6 of IC204 cannot be set to 600 mV with R255, suspect delay line X204.

Intermittent colour flashing on replay of own recordings No pilot burst being recorded due to faulty IC202.

V9600:

Intermittent unloading Check loading motor belt; if OK, replace motor.

Poor rewind Fit modified upper cylinder, replace rewind idler pulley assembly and clean all drive surfaces. If necessary replace head disc.

Jerky or inconsistent take-up torque Clean or replace idler pulley. If fault persists replace the motor.

Mechanical noise on playback, threading and eject operations rapid 14 V power rail up to 22 V due to faulty Q902.

Cassette compartment will not insert or eject Check QL83/84/85/86, mounted on subpanel to the rear of the cassette compartment.

Drum motor continues running after eject Check pin 7 of IC601; if there is no change of state on eject, replace IC601.

Hum bars or thin lines moving vertically on picture Add capacitor 0.1 μF from Q008 base to tuner chassis.

No flashing digits when setting timer Faulty timer IC.

Appendix 1

AV SOCKET TYPES

With the increasing availability of TVs and other equipment designed to work with base band audio and video signals, a worthwhile improvement in performance can be gained by using direct audio and video links between VCR and peripheral gear. Some videorecorder models of Philips and allied origin are fitted with an exclusive 20-pin AV socket, for which a special adaptor is required. Some small portable machines have conventional output sockets, but require all inputs to be routed via their camera sockets; to use other signal sources, plug-adaptors will be required, usually incorporating an attenuator in the audio signal path.

The following list of videorecorder makes and models shows the type of socket fitted to each, and should be used in conjunction with the diagrams below, in which each socket type is drawn, with the position and function of each pin.

The 5-pin DIN socket (1) is concerned only with audio signals, and is compatible with the 7-pin system (3) which also has the facility (pin 6) to transfer remote control data. The 6-pin DIN socket (2) may be found on TVs and VCRs (notably Grundig types and Japanese models made for use in Europe) and caters for incoming or outgoing signals according to the switching voltage on its pin 1. The 8-pin socket (4) is fitted to some European-market TV sets, where it acts as an input route only. The Scart connector (5) may also be referred to as **Euroconnector** or **Peritel**. It has comprehensive in/out facilities and is fitted to several late models of TV and VCR; in the latter case only audio and composite video connections are used.

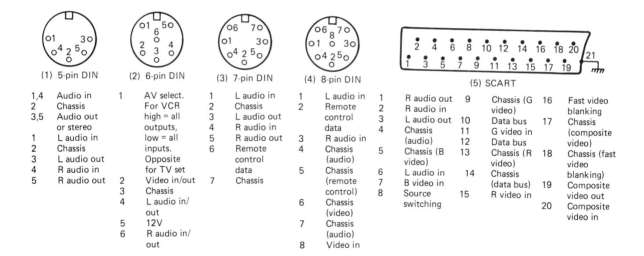

	(1) 5-pin DIN		(2) 6-pin DIN		(3) 7-pin DIN		(4) 8-pin DIN		(5) SCART			
1,4	Audio in	1	AV select.	1	L audio in	1	L audio in	1	R audio out	9	Chassis (G	16	Fast video
2	Chassis		For VCR	2	Chassis	2	Remote	2	R audio in		video)		blanking
3,5	Audio out		high = all	3	L audio out		control	3	L audio out	10	Data bus	17	Chassis
	or stereo		outputs,	4	R audio in		data	4	Chassis	11	G video in		(composite
1	L audio in		low = all	5	R audio out	3	R audio in		(audio)	12	Data bus		video)
2	Chassis		inputs.	6	Remote	4	Chassis	5	Chassis (B	13	Chassis (R	18	Chassis (fast
3	L audio out		Opposite		control		(audio)		video)		video)		video
4	R audio in		for TV set		data	5	Chassis	6	L audio in	14	Chassis		blanking)
5	R audio out	2	Video in/out	7	Chassis		(remote	7	B video in		(data bus)	19	Composite
		3	Chassis				control)	8	Source	15	R video in		video out
		4	L audio in/			6	Chassis		switching			20	Composite
			out				(video)						video in
		5	12V			7	Chassis						
		6	R audio in/				(audio)						
			out			8	Video in						

VCR socket codes:

	Video	Audio
A	BNC	3.5 mm jack
B	Phono	5-pin DIN
C	BNC	5-pin DIN
D	PL259	5-pin DIN
E	BNC	RCA phono
F (note 6)	6-pin DIN	6-pin DIN
G	RCA phono	RCA phono
H	PL259	RCA phono
J (note 7)	BNC	7-pin DIN
K	SCART	SCART.

Notes: (1) 20-pin socket; special adaptor needed for AV operation.
(2) Playback only.
(3) UHF interface only.
(4) Input only via camera socket.
(5) Late models group K.
(6) Many group F VCRs have additional 5- or 7-pin audio sockets.
(7) Group J is compatible with group C.
(8) The JVC HRD 725EG is fitted with RCA phono sockets for audio, in addition to its SCART connector.

Akai

VP88	B
VP7100	A
VS1	G
VS2	B
VS4	G
VS5	B
VS6	G
VS8	G
VS10	C
VS77	A
VS9300	D
VS9500	D
VS9700	B
VS9800	D

Amstrad

7000	E

Baird

8900	D
8902	D
8903	D
8904	D
8906	A
8922	D
8924	C
8925	A
8930	C
8940	C
8941	J
8942	J
8943	J
8944	J

Beocord

8800V	(1)
8802V	(1)
VHS80	F
VHS90	K

Blaupunkt

RTV100	E
RTV200	E
RTV202	E
RTV211	F
RTV222	E
RTV224	E
RTV301	F
RTV321	F
RTV322	F
RTV358	F
RTX100	F
RTX200	F
RTX250	F

Bosh-Bauer

VRP20	F

Bush

BV6900	G

Decca

VRH8300	C
VRH8400	J
VRH8500	J

Fidelity

VTR1000	G
VTR1001	G

Ferguson

3292	D
3V00	D
3V01	A
3V16	D
3V22	D
3V23	C
3V24	A
3V29	C
3V30	C
3V31	J
3V32	J
3V35	J
3V36	J
3V38	J
3V40	A (4)

Fisher

FVHP520	G
FVHP530	G
FVHP615	E
FVHP715	E
VBR330	G
VBS7000	D
VBS7500	E
VBS7600	E
VBS9000	H
VBS9900	E

GEC

V4000	G
V4001	G
V4002	G
V4004	G
V4100	G

Granada

CHA VH2	G
VHS TJ1	D
VHS VH1	G
VHS VH2	G
VHS WH1	G
VHS WH3	G

Grundig

2×4GB	F
2×4Plus	F
2×4Super	F
1600	F (2)
2080	F
SVR4004	(3)
SVR4004AV	F
SVR4004EL	(3)
SVR4004EL AV	F
VP100	F
VS180	F
VS200	F
VS220	F

Hitachi

VT7	G
VT11	G
VT17	G
VT19	G
VT33	G
VT35	G
VT57	G
VT88	G
VT3000E	D
VT5000E	G
VT5000ER	G
VT5500E	G
VT5550	G
VT6500	G
VT7000	G
VT8000	G
VT8040	G
VT8300	G
VT8500	G
VT8700	G
VT9300	G
VT9500	G
VT9700	G
VT9800	D

ITT

TR3913	C
TR3943	C
TR3984	C
VR482	(1)
VR483	(1)
VR580	(1)
VR3605	C
VR3905	C
VR3919	C
VR3975	C
VR4812	F
VR4833	F
VR4913	F
VR4963	F

JVC

HR2200	A
HR2650	C
HR3200	D
HR3300	D
HR3320	D
HR3330	D
HR3360	D
HR3660	D
HR3860	D
HR4100	A
HR7200	C
HR7300	C
HR7350	C
HR7600	C
HR7650	C
HR7655EG	F
HR7655	C
HR7700	C
HRC3	A (4)
HRD120	J
HRD225	J
HRD 725EK	E
HRD 725EG	K (8)

Korting

VR8	F
VR10	F

Loewe

OC6023	F

Luxor

9213	C
9214	C
9221	C

Marantz

MVR500	F

Metz

VR9913	F

Mitsubishi

HS200	D
HS210	D
HS300	C
HS301	C
HS302	C
HS303	C
HS304	B
HS306	B
HS307	B
HS310	C
HS320	C
HS330	B
HS400	G
HS700	B
HS2006	D

Murphy

MVR7007R	E

NEC

PVC470E	E
PVC2400	E

Nordmende

Spectra VIS	D
Spectra W	D
V100E	C
V1001	C
V100H	F
V100K	F
V101	C (5)
V102	C (5)
V150	F (4)
V200	D
V250	C
V300E	C
V300H	F
V300K	F
V302	F (5)
V303	C
V350	F
V380	F
V400	D
V460H	D
V500E	C
V500H	F
V500K	F

Panasonic				
NV100	E			
NV200	E			
NV333	E			
NV366	E			
NV370	E			
NV688	E			
NV730	E			
NV777	E			
NV788	E			
NV830	E			
NV850	E or F			
NV2000	E			
NV2010	E			
NV3000	E			
NV7000	E			
NV7200	E			
NV7800	E or F			
NV8170	E			
NV8600	G			
NV8610	E			

Philips

| N1500 | (3) |
| N1700 | (3) |

VR2020 (1)

VR2020	(1)
VR2021	(1)
VR2022	(1)
VR2023	F
VR2024	F
VR2025	F
VR2220	F
VR2323	F
VR2324	K
VR2334	K
VR6520	C

Pye

20VR20	(1)
20VR22	(1)
20VR23	F
22VR20	F
23VR24	K
23VR33	K
6520	K

RCA

| TC 3451X | E |

RANK

| 6910 | G |

Saba

VR2000	F
VR6000	F
VR6004	K
VR6005	K
VR6010	F
VR6012	F
VR6020	F
VR6022	F
VR6024	F
VR6028	F
VR6068	F
VR6069	F
VR7720	F

Saisho

| VR705 | E |

Salora

| SV8300 | B |

Sansui

| SV7020 | C |

Sanyo

VC8700	D
VC9700	D
VPR5800	E
VTC3000	E
VTC3000P	A
VTC5000	E
VTC5150	E
VTC5300	E
VTC5400	E
VTC5500	H
VTC5600	H
VTC6500	H
VTC9300P	D
VTC9300PN	D
VTC M10	E

Sharp

VC220	G
VC381	C
VC386	C
VC387	C
VC388	C
VC390	C
VC482	C

Sony

SL3000	E
SL8000	E
SL8080	E
SLC5	E
SLC6	E
SLC7	E
SLC9	E
SLF1	E

VC483	C
VC486	C
VC496	C
VC2300	B
VC3300	C
VC6300	D
VC7300	D
VC7700	D
VC7750	D
VC8300	D
VC9300	C
VC9500	C
VC9700	C

SLC20	E
SLC30	E
SLC40	E
HFL100	E

Tatung

| VRH8300 | C |
| VRH8500 | J |

Teleton

VG200B	G
VG220B	G
VG240B	G

Toshiba

V31B	G
V33B	G
V55B	J
V57B	J
V5470	G
V8600	G
V8650	G
V8700	G
V9600	G

Appendix 2

GLOSSARY OF VIDEO TERMS

AC motor drive Instantaneous motive power and long term reliability are required of the drive system. The AC motor drive system satisfies these requirements. It is quiet and reliable, ensuring a clear stable TV picture.

ACC (automatic colour control) This is the name given to a circuit which maintains constant amplitude level of the colour burst and hence the colour signal against input variations.

AFC (automatic frequency control) Although the TV horizontal oscillator circuit is of the self-oscillating type, increased stability of operation is achieved by triggering this oscillator circuit by means of the sync signal (15.625 kHz) separated from the composite video signal. However if the horizontal sync signal separated from the composite video signal alone is used to directly trigger the oscillator circuit, noise components which are within the sync signal's frequency band will cause non-uniformity in the oscillator cycle with a corresponding degradation in picture stability. Therefore an additional circuit is placed within the horizontal oscillator circuit for the purpose of detecting phase differences between the sync signal and oscillator frequency and automatically governing the oscillator. This is called the AFC (automatic frequency control) circuit.

AFC (colour signal) As applied to a video-recorder this is the term given to the $160 f_h$ oscillator circuit in the colour system when it is locked to video syncs to stabilise the frequency.

AFT (automatic fine tuning) Used to prevent tuning drift of received transmissions or stabilisation of the exact tuning point.

After image When panning with a video camera, or when the subject moves rapidly, the image of the subject may streak or smear on the screen. This after image is a greater problem at low light levels.

AGC (automatic gain control) AGC is used to maintain a fixed output signal level by minimising the effects of input level fluctuations. When the amp detects a change in input signal amplitude, this is sent back through a feedback circuit as an automatic gain control signal and the fluctuation is compensated for.

ALC (automatic level control) Usually applied to audio recording circuits but can be used instead of AGC as a term for the automatic level control of recording video signals.

Amount of head protrusion For maximum tape-to-head contact, the heads protrude slightly from the drum. This allows them to actually penetrate the tape to a minor degree (head penetration). Excessive head penetration may cause tape stretch and head wear.

Angle adjustment In a two head VTR it is necessary to adjust the head gaps so that they are 180° from the centre of the drum. If this angle is incorrect, tape compatibility suffers. Proper adjustment is achieved by playing back a test tape and adjusting the screw of head B while using head A as a reference. The adjustment screw used is a special kind with very fine pitch.

APS (automatic programme search) An electronic system for finding the start or ending of a recording.

Aperture correction When an iris is opened from the 'pin hole' or closed position some de-focussing of the image occurs. Aperture correction is the term given to an electronic method of sharpening the image to correct for this. It can be applied to any general procedure for enhancing the sharpness of a video signal.

Assemble edit Electronic method for joining two or more sequential recordings together without intermediate picture disturbances.

Audio dub Electronic method or recording sound onto tape with a previous visual recording and without disturbing it.

Auto colour balance The ability of a camera to automatically adjust to ambient light colour temperature so that a white image has minimum colour component.

Auto focus The ability of a camera to focus onto image targets by its own control systems.

Auto rewind A microcomputer function to re-wind a videocassette when the end of tape sensor is activated in record, some models will provide the same function in replay.

AUX (auxiliary) An additional input terminal found on amplifiers and used for monitoring or amplifying inputs that do not require equalisation. Sometimes called a 'line in' terminal.

AV (audio visual) The name given to audio and video equipment, also refers to baseband inputs on video equipment or TVs.

Azimuth recording The video signal recording method used in domestic VTRs. It takes advantage of azimuth loss which occurs particularly at higher frequencies when the azimuth angle of the record-ing head gap is different from that of the playback head gap. By giving video heads A and B a relative angle difference, when the track recorded by head A is played back, the track recorded by head B will not be reproduced because of azimuth loss. In the same way, during track B playback, the track A signal will not be picked up. This allows extended playing and recording time since the empty space between tracks is not needed for preventing cross-talk. Diagonal head gaps are used for this kind of azimuth recording.

Back focus Adjustment between the lens and optical pick up device to focus the lens, at infinity or wide angle, on the photosensitive surface.

Back tension The tension in gm.cm (gm × cm) in a video tape ribbon between the supply spool and the video head drum assembly.

Base Band *see* CVBS.

Beam limiting In video cameras with vidicon types of pick up tubes the electron beam has to be limited to prevent internal electron defocussing, the limiting level is approximately 1.5 V pp at the video output.

Beta format An inclusive term describing the home video cassette system developed by Sony.

Black level Corresponds to brightness on a TV receiver. In a camera it refers to the DC level of optical black, or pedestal level.

Blanking TV scanning goes from left to right and top to bottom. But when the scanning beam gets all the way to the right (or bottom) it must go back to the left (or top). This retrace time is called the blanking period. In order to maintain good picture quality, the electron beam is shut off during the instant it takes to get to the beginning of the next scanning line.

BNC (British Naval Connector) The name given to the standard bayonet type of video connector terminal.

Booster An amplifier used to increase signal strength particularly between an antenna and a TV receiver.

Burn Permanent or semi permanent image on the camera pick up tube caused by excessive bright light source. It can often be healed by long term exposure to a white image.

Burst flag The term given to a pulse on the R-Y and B-Y signals in an encoder which gate the subcarrier to form the colour burst pulse.

Burst signal Also called the colour sync signal, the colour burst consists of 10 cycles of the chrominance subcarrier (4.433619 MHz) added at the end of the horizontal sync signal of the TV signal. This burst signal indicates the frequency and phase of the chrominance subcarrier being transmitted and it becomes the reference for phase comparison during reception (amplitude for chro-ma, phase for hue determination) in generating the continuous carrier necessary at the time of deriv-ing the colour signal.

Cable compensation unit Signal loss and distor-tion will result if the video signal is sent through a long cable. This type of unit is used at the receiving end of the cable to return the signal to its original level and waveform.

Camera switcher A piece of video equipment used in a multi-camera system to switch between cameras. A switching system may provide such effects as the lap (as one picture fades out, another fades in), fading (fading a single picture in or out), superimposing, wiping and so forth.

Capstan servo This is a servo system that maintains proper capstan motor rotational speed during recording by comparing the phase of the capstan motor PG with that of an external signal from the video input terminal, or with the internal sync signal output when there is no input from outside. During playback, proper rotational speed of the capstan motor is maintained by comparing the CTL signal phase with that of an external input or the internally generated sync signal.

Carrier balance The controls that balance colour encoder modulators to eliminate residual carrier at the output of the encoder.

Carrier leak Although the video signal is recorded on the VTR as an FM signal, if a plain carrier signal appears during playback, it will cause a beat effect appearing on the screen as a series of wavy lines. This phenomenon is called carrier leak.

CATV Community antenna television. A system employing a single large antenna having very good performance to supply a community with high quality reception via cable.

CCD (charged coupled device) A matrix of photoconductors which make up a solid state camera tube.

C-format Small VHS format.

CCIR (International Radio Consultative Committee) This international committee investigates various problems concerning radio and TV broadcasting and makes suggestions for solutions.

CCTV (closed circuit TV) A non-broadcast system employing cables to connect cameras or VTRs with a specific building, factory, school, and so on.

Chroma (degree of saturation) Colours are often described as, for example, dark reds and light reds. In technical terms, a dark red is a saturated colour.

As white light is added, it will become less saturated and will be seen as a lighter red.

Chroma refers to this degree of saturation (regardless of hue). On a TV monitor, chroma is adjusted by the COLOUR control knob.

Chrominance signal Both hue and chroma are combined in the chrominance signal. (Compare with luminance.)

Clogging When the gaps in the heads become clogged with dirt or by abrasion of the tape surface, it will result in noise and inability of the VTR to record and play back properly. If the CTL head becomes clogged then the servo will cease to operate. If the heads are dirty, they must be cleaned.

Close-up A type of shot in which a person's head and shoulders, a hand, feet or other objects fill the frame.

C-mount A standard type of lens mount used on 16 mm movie cameras. It is a screw-in mount and is often found on video cameras also.

Coaxial cable A type of cable commonly used in video because of its suitability for transmitting video and RF signal. Coaxial cable has an inner conductor that is coaxial with a braided sheath used as the outer conductor. The two are separated by a dielectric. Coaxial cable is cylindrical in shape.

Colour bar Within a colour test signal, the colour bar contains bars of colour and peak black and white (from left to right: white, yellow, cyan, green, magenta, red, blue) used to gauge luminance and chrominance of the picture.

Colour filter Orange coloured filter in a camera lens which reduces blue light falling on the pick up tube for outdoor colour balance.

Colour signal The composite colour signal consisting of the luminance signal conveying brightness, the chrominance subcarrier signal conveying hue and saturation, and the colour sync signal (colour burst signal). The term may also refer to the chrominance signal alone.

Colour temperature When a black body is heated to a high temperature, it emits light. The relationship between the colour of the light and the temperature is constant, thereby allowing colour to be expressed in terms of absolute temperature. Reddish colours have a lower colour; bluish colours have a higher colour temperature.

Compatibility   If a tape made on one VTR can be played back on another and vice versa, then the tape is said to be compatible with both units.

Complementary colours   When light of one of the primary colours (red, green, blue) is subtracted from white light the complementary colour (cyan, magenta, yellow, respectively) is produced.

Composite sync signal   A sync signal composed of the horizontal sync signal, vertical sync signal, and equalising pulse signal. The composite video signal is produced by adding this to the video signal. If the signal is colour, the colour burst is included in the composite sync signal.

Composite video signal   The combination of video signal, blanking signal, and composite sync signal. Also known as CVBS signal where colour is carried.

Condensation   Moisture in the air will form water droplets on a cold piece of glass or metal in a warm room. This same kind of condensation will occur on the head drum under certain conditions. If a VTR is used when there is condensation on the drum, the tape will stick to the drum and be ruined.

Control console   A control console is used within the control room of a TV studio to give centralised control of all operations. It has facilities for switching and monitoring cameras, VTRs, broadcasts and film chains, along with sound mixing.

Control signal   This is necessary to the servo circuit used to maintain proper head rotation. During recording the vertical sync signal is separated from the video input signal and recorded on a separate track (CTL track). This is picked up on playback and becomes the reference signal for servo system operation.

Control track   The track used to record the control signal. The video and audio signals are also recorded on other tracks.

Crosstalk   Unwanted spillover of signal between one recorded video track and another.

CTL head   Records and plays back the video control (CTL) signal. The CTL head and audio head are a single unit.

Cue Review   Also called search or shuttle mode. Video playback at 11 times normal speed, forward and reverse (on average).

CVBS (composite video burst (or blanking) and syncs)   A complete video signal, referred to a baseband video.

Dark current   The very low residual vidicon tube beam current which still flows at black level, kept to a minimum to reduce picture noise.

Depth of field   The amount of distance between near and far objects which are both in focus.

Dolby   A patented method of sound noise reduction used in audio and VHS-video machines.

Dropout   Transient loss or reduction in the playback signal due to imperfections in the tape itself. Although this often results from such causes as old tape, inferior quality tape, and improper storage and maintenance of the tape and machine, even the best quality tape contains a certain percentage of dropouts, due to limitations in tape manufacturing technology. The dropout compensator circuit functions to prevent these inherent defects from visibly affecting the picture.

Drum servo   This servo system governs head rotation by comparing the phase of the vertical sync signal derived from the input video signal with that of the 25 PG signal produced by drum rotation. Any difference between the two is detected, the error voltage (DC) amplified and applied to the drum motor.

Dubbing   After the video signal has been recorded on the tape, sound effects and dialogue are added while watching the playback image.

Dynamic focus   Modulation of camera tube scanning waveforms to reduce defocussing around the edges and prevent green tinting.

Dynamic shading   Horizontal and vertical sawtooth and parabolic waveforms used in the red and blue gain controls to improve camera tube 'purity' and prevent tinting around the picture.

Editing The process of rearranging, adding, and removing sections of the picture and sound already recorded on tapes. Two different modes of electronic editing are used with video tape; 'insert' editing means putting new picture and sound in the middle of a recorded section. Assemble editing is the technique of adding on different picture and sound from the middle of a recorded tape to the end.

E-E (Electric-to-Electric) When the VTR record button is depressed, the record output circuit is connected to the playback input circuit so that the video signal to be recorded can be monitored on a TV set. Since the magnetic components (heads and tape) have nothing to do with this signal and it is just passed as is from one circuit to another, it is called the E-E mode.

EIAJ Electronic Industries Association of Japan.

End sensor This is an auto-stop device at each end of the tape. It prevents the leader tape from contacting the video heads thus protecting them from possible damage.

Equalising pulses A series of six pulses inserted before and after the vertical sync signal to ensure proper interlaced scanning.

EVF (electronic viewfinder) Camera viewfinder monitor.

External sync In addition to the signal produced by the camera tube by converting the subject image from the lens into current waves, the video signal also needs vertical and horizontal sync signals to make up a picture. This sync may be fed to the camera from outside or generated by an oscillator in the camera. The former is called external sync, the latter internal sync.

Fade in The technique of gradually increasing the strength of the video signal so that it fades in on the screen finally becoming normal picture.

Fade out The technique of gradually fading out a picture until finally no picture appears on the screen.

Field and frame In the system used in Europe there are 50 fields and 15 625 horizontal scanning lines (625 lines × 25 frames) per second used to break down the image (in a camera) or reconstruct it (in a receiver). Interlaced scanning is employed to reduce flicker. This involves scanning half (312.5 lines) the lines (every other line) during one vertical scan and then scanning those lines in between the first lines during the following vertical scan from top to bottom of the screen (another 312.5 lines). Two vertical scans results in scanning of all the horizontal scanning lines (312.5 × 2 = 625 lines) and the completion of one picture. In other words since it takes two vertical scans to complete a picture, 25 pictures can be broken down or reconstructed every second. The term 'one field' refers to the rough picture described during one vertical scanning period ($^1/_{50}$ second). 'One frame' is the complete picture produced by two vertical scans ($^1/_{25}$ second = 2 fields).

FM recording Direct recording of the video signal is difficult since it covers a broad range from DC to 4 MHz. Therefore a carrier wave produced by an oscillator in the VTR is frequency modulated (FM) in accordance with the amplitude in the video signal. The FM signal thus produced is recorded on the tape instead of the original video signal. This FM recording method has the advantages of permitting recording down to DC (0 Hz), and eliminating level fluctuation by means of a limiter circuit.

Focal length The distance from the optical centre of a lens to the principal focus (the point where parallel rays of light focus).

Full logic transport control Circuitry that provides smooth ordered logic switching between modes (forward, rewind, fast forward, and stop). Commonly used in both audio and video tape recorders, this type of mode switching system makes possible remote control automatic electronic editing.

Ghost An additional vague image appearing to the right of the main image on a TV screen. Such ghosts are caused when the antenna picks up the same signal arriving by slightly different routes (multipath reception). It may also arise internally when impedance mismatching results in the generation of a reflected wave.

Guard band The empty space between video tracks provided to prevent crosstalk. There are no guard bands in the current domestic VTR systems since azimuth recording renders them unnecessary for the achievement of a good picture.

Head gap   The tiny space between the two pole pieces of the ferrite core used in the video heads. Typical head gaps are $0.5\,\mu m$ wide and are filled with special glass which maintains constant dimensions.

Head to tape speed   Since the video signal reaches such high frequencies, it cannot be recorded at ordinary audio tape speeds. However, the relative speed can be greatly increased by using a rotating head.

Helical scanning   One type of video tape recording. As the heads rotate in the horizontal plane, the tape passes over the drum diagonally resulting in diagonal tracks on the tape. The term 'helical' is derived from the shape the tape path describes as it is wrapped around the cylindrical drum. Helical scanning permits both mechanical and electrical simplification of construction and makes possible simulated slow motion and still playback. Maintenance is also easier with helical scan machines.

Hue   Refers to the characteristic of a colour that distinguishes it as red, yellow, green, blue and so forth. In other words, hue is determined by the dominant wavelength regardless of saturation.

Impedance matching   The process of matching the impedance of the source and load for the best transmission of the signal. Cables and connectors can also be thought of as a kind of impedance. The coaxial cables used for connecting video equipment usually have an impedance of $75\,\Omega$. This is why the standard video terminal impedance for video rated equipment is also $75\,\Omega$.

Impedance matching box   A device used for matching impedance between the antenna and coaxial cables or between coaxial cables and the receiver terminal.

Insert edit   An elaborate technique for inserting new audio or video material into previously recorded material without disturbance at the start or ending of the insert.

Interlace   A scanning system in which the lines of each frame are divided up into two fields taking up the even and odd scanning lines of the raster which are scanned alternately.

Interlaced scanning   In the PAL system there are in total 625 scanning lines. These are not scanned sequentially; rather first the 312½ odd lines are scanned and then the even ones (312½). The combined result of these two fields is one complete frame. This method of interlaced scanning reduces flicker without the need for broadening the video signal band (as would be the case if sequential scanning were used). *See also* Field *and* Frame.

Internal sync   Cameras equipped with a built-in oscillator that generates the sync signals are said to have internal sync. This is standard practice in domestic cameras.

Jitter   Instability in the playback signal due to tape or head speed fluctuations.

Jog   Special playback modes, such as slow, still, frame by frame, and three times normal speed.

Line in/line out   The terminals which apply signals strong enough to be supplied to power amplifiers by amplifying weak signals from a microphone amp or head amp), or sometimes improving frequency response correcting electrical features (by the equaliser circuit), are called line out terminals. The terminals of amplifiers which amplify signals from tape recorder or tuner output terminals are called line-in terminals.

Loading   In the cassette format, an automatic loading system is needed to take the tape out of the cassette and wrap it around the head drum and capstan so that recording and playback may take place. The term 'loading' refers to the entire mechanical operation from cassette insertion to completion of tape path set-up. 'Unloading' refers to the opposite operation during which the tape is put back into the cassette.

Luminance signal   In contrast to the chrominance signal, the luminance signal conveys the information on picture brightness (tone). Only the luminance signal is transmitted in the case of black and white broadcasts.

LUX   The measurement of light intensity falling upon an area. The light given out by 1 candle one foot away is 1 Footcandle $= 10.76\,Lux$.

M-loading   Standard threading system used in VHS machines. The tape is wrapped round the head in M-configuration, making for simple threading mechanisms and short loading time.

Macro   Special adaption to a lens for very close up focussing.

MAG MAG Tube   Camera vidicon tube with magnetic deflection and magnetic focus (focus coil).

ND filter   A neutral density filter. An ND filter is an optical filter used on cameras. As opposed to colour correction filters, an ND filter passes all colours (frequencies) equally. An ND4 passes ¼ the light, an ND6 passes ⅙ and so forth. An ND filter is used instead of a small aperture in bright conditions to reduce the depth of field.

Ni cad (nickel cadmium) batteries   Have a very low internal resistance and can be rapidly charged although some heat is generated, very dangerous if short circuited.

Noise canceller circuit   The video signal band includes harmonics of the sync and luminance signals. This causes cross-colour noise and must be removed to obtain good picture quality. In some machines this is achieved by separating the signal and noise components using a high filter and limiter and then applying the noise component with equal amplitude and reverse phase back onto the signal. This results in cancellation of the noise.

NTSC   (National Television System Committee)   A group of TV industry specialists who developed the present colour telecasting system used in the US, Canada and Japan.

Open-reel   A tape recorder employing separate reels. Domestic VTRs of all formats use cassettes, in which the tape is fully enclosed.

Overlap   The recorded video signal is reproduced by the alternating output of two video heads but the beginning of head B output is allowed to overlap a bit with the end of the head A output in order to eliminate blank areas. The amount of this overlap is between 3 H and 12 H. The term may also be used to refer to the lap dissolve technique of switching pictures by reducing one while increasing the other.

Overmodulation   Overmodulation occurs when a large input increases the oscillator frequency in the FM modulation circuit. This may appear as a black edgelike effect over the right side of white portions of the monitor screen.

Ω type   Looking straight down on the drum of a helical scan VTR, one can see that the tape is wrapped about 180° around the drum. This name is derived from this shape.

PAL (phase alternation line)   The colour TV system used in West Germany, Holland, Britain, Switzerland and other countries.

Panning   Pivoting the camera in the horizontal plane.

Pause   Stopping a tape transport while remaining in the same mode (playback or recording). In contrast to using the stop button (or lever), one can go from pause back into the record mode directly without having to press the record button again. This is handy during editing and for eliminating undesired sections of a programme during recording.

PG (pulse generator)   In the VTR, there are two coils around cores (PG coils) 180° apart on the head drum and there are magnets or iron pieces affixed to the video head disc holding the video heads. When the magnets (or iron pieces) pass the coil, the pulse produced is positive on the approach and negative as the magnet leaves. One pulse is generated with each video head rotation and the signals from the two PG outputs are formed into a rectangular wave. The pulse signal obtained is used as the reference signal during playback and as a comparison signal during recording.

Phono   Name for the RCA phono connector used widely on Japanese audio equipment and for the audio connector of video equipment.

Photoconductive layer   The light passing through the lens is focused to form an image on the photoconductive layer on the target with the camera. Resistance varies as light strikes the target and the resulting signal reflects the level of light at each point in the image.

R/B carrier   The component obtained by scanning the stripes on a colour camera tube as a high frequency carrier containing the colour information.

Resolution   Ability to reproduce subject detail. A pattern of black and white lines is reproduced on the screen and the number of lines visible is used to express the degree of resolution of the equipment.

Return video Sending the processed video signal back from the control room to the camera where it can be monitored on the viewfinder by the camera operator.

Rewind The tape transport mode in which the tape is wound at high speed onto the supply reel. Except for the V2000 format, video tapes are only played on one side (in contrast to audio cassettes). Therefore they must be rewound in order to be viewed over again from the beginning. However, one big advantage of the cassette system is that rewinding is not necessary before ejecting the tape. (It is necessary to rewind the tape on an open reel machine since the reels are separate and not contained within a single cassette box.)

RF signal Any frequency at which coherent electromagnetic radiation of energy is possible.

RF unit A device that permits a conventional TV receiver to be used as a monitor for a VTR. The RF unit converts the sound and picture signals from the VTR into the same kind of TV signal broadcast by TV stations. Thus by merely setting the TV to the prescribed channel, one can view the VTR playback picture or that of a video camera in the same way as watching an ordinary TV broadcast.

RF transmission method A method of transmitting the video signal. By converting a VTR or video camera video output signal into the same kind of TV signal and sending it through a cable, one can view the picture on a TV connected to the other end. If the converted signal frequency is matched to that of an inactive TV channel then both ordinary channels and the VTR signal can be selected by just turning the channel knob. A splitter may also be used to connect a number of TV sets for simultaneous reception. This type of video transmisson is called the RF method or community TV system.

Rotary transformer The transformer that provides the signal current to the rotating video heads during recording, and picks up the signal from the heads during playback.

Rotating drum In helical scan VTRs, the tape is wrapped around the rotating drum diagonally (it forms a helix), so that the rotating heads can trace the tape surface at the correct diagonal angle. In contrast to conventional drums, that used in the

Betamax has three sections. The upper and lower layers are stationary, while the video heads are affixed to the centre section and rotate as one body. The centre drum and the heads thus rotate as one.

Rotating head At least 200 times the normal audio tape speed is required to record the high frequencies of the video signal. This speed is achieved by rotating the heads while maintaining a slow tape speed, thus achieving a high relative speed. In a helical scan system, the heads rotate in a horizontal plane as the tape passes around the drum diagonally. The tracks traced on the tape are therefore also diagonal. Compared to audio recording, this technique greatly increases the recording area in terms of efficient use of the tape surface.

S-VHS A high band version of the VHS system, not compatible with standard VHS.

S-terminal Name given to the connections where the video signal is split as two signals, luminance and chrominance. Used for S-VHS.

Saticon Trade name for a variation of the vidicon camera pick up tube which uses Selenium-Arsenic-Tellurium chalcogenIde and the photo-CONversion layer. In common use for video cameras and camcorders.

Scanning In television, an entire picture is never sent at one time. Rather, it is broken up into discrete lines and the information regarding each line's brightness (tone) and colour (hue and chroma) is sent separately and sequentially. The receiver picks this up and reconstructs the picture. The process of breaking down and building up the image is called scanning. Ordinarily the picture is scanned from left to right (horizontal scanning) and from top to bottom (vertical scanning). These two types of scanning are carried out simultaneously.

SCART 21 pin euroconnector, supersedes DIN for AV usage also known as 'peritel' (peripheral television).

Search *See* Cue/Review.

SECAM A colour television system used in France, the Soviet Union and some other countries.

Separate mesh   In some vidicon camera tubes, a wire mesh electrode is employed in addition to the focusing electrode. The voltage of the former is made 1.6–1.7 times that of the latter. Separate mesh vidicons have improved resolution.

Servo   A servo mechanism is a type of automatic governing device used to maintain a fixed speed, position, or angle of operation.

Shading   Although the camera should produce a picture that is a replica of the brightness gradients in the subject, sometimes 'shadows' appear that are caused by the camera alone. This phenomenon is called shading.

Shuttle   *See* Cue/Review.

Signal splitter   A device that distributes a single signal equally to two or more other lines. Signal strength drops in proportion to the number of lines to which the signal is distributed (distribution loss) (4 dB loss for a 2-way distributor; 7–8 dB for a 4-way). If distributor terminals are not being used they should be connected to a dummy load.

Simultaneous recording   Recording picture and sound at the same time (on the same tape).

Skew   When back tension on the tape is not correct, the upper part of the reproduced picture will be bent out of shape. This condition is called skew.

Slave TV (slave monitor)   In a system employing several monitors to view the same source, those monitors without selectable reception capability are called slaves and the main unit which does have a means to select between different inputs and receive various programs is called the master. Thus the picture and sound reproduced by a slave is dependent upon the signal sent to it (via cable) by the master.

Slip ring   Used to make connection to rotating members. In domestic VTRs slip rings are used in some drive motors, and to convey operating voltage to the DTF tracking bars in V2000-format machines.

Slow motion   Movement of subjects in picture at slower than normal speed.

S/N ratio   The ratio between the desired signal and the unwanted noise components. The larger the S/N ratio figure, the clearer the picture and sound.

Splicing tape   Special tape applied to the back of two pieces of tape that have been matched up. Splicing tape is rarely used in video because of the danger of harming the heads and the unstable image that appears at the point of the splice.

SSG (sync signal generator)   IC or circuit which produces the scanning drive pulses, blanking, syncs and subcarrier.

STAMAG   A vidicon tube with static focus within the electron gun and magnetic deflection, such as the Saticon.

Standard lens   The lens considered standard for a video camera depends on the size of the camera tube. For a 1-inch vidicon the focal length is 25 mm, for a ⅔ inch vidicon the focal length for a standard lens would be 16 mm. If a lens is longer than the standard focal length for a certain size camera tube, it is called a telephoto lens; if it is shorter than the standard, it is called a wide-angle lens. A zoom lens offers continuously variable focal length from wide-angle to telephoto without affecting focus or aperture.

Static shading   Changes in black level due to uneven dark current or target surface sensitivity at or around the edges of the pick up tube with no illumination.

Streaking   The name given to the phenomenon of a black or white horizontal line following a bright highlight due to poor bandwidth or clamping.

Subcarrier phase shifter   The colour burst determines the hue of colour produced in a colour television. But in a multicamera system the lengths of cables connecting each camera with the switcher will usually differ and cause phase shift. Then when the switcher changes from one camera to another shooting the same subject, the hue will not be the same. Therefore a subcarrier phase shifter is employed to adjust the phase of the colour burst to compensate for the different cable lengths thus ensuring that the hue of the picture received from each camera is the same.

Superimpose A video technique by which lettering or a chart can be reproduced on top of another picture. Two cameras are used. For example, while camera A shoots one scene, camera B shoots a caption written in black and white. By electronically processing the signal from camera B, just the black or just the white portions can be removed and the remainder superimposed on the picture from camera A.

Sync generator The sync signal required by the video camera may be either internally or externally generated. In the latter case it uses external sync supplied by a sync generator (which may be in the VTR, switcher, SEG, or camera control unit). The camera will not operate properly and a stable recording cannot be made unless sync is supplied.

Sync signal The pulse signals used to synchronise the operation of video equipment. The horizontal sync signal determines the start of horizontal scanning; the vertical sync signal determines the timing of the beginning of vertical scanning.

Syscon Nomenclature for the microcomputer systems control circuits for video recorder functions and tape control.

Synthesised vertical sync pulse A vertical sync pulse inserted into the replayed video signal to stabilise the TV scanning circuits during visual search and still picture modes of operation.

Tally light A light on the video camera that tells the operator that the signal from his camera is being used. Two lights may be provided: one on top of the camera, the other within or next to the viewfinder.

Tape copying Video tape can be copied in the same way as audio tape by using two VTRs. There is of course some inevitable degeneration of picture quality.

Tape counter The figures on the tape counter advance as the tape is used. It may be used for finding a specific place on a tape or judging the amount of tape remaining.

Tape guide Used to guide the tape properly past the video heads, CTL head, capstan, and other areas along the tape path. They usually consist of metal cylinders (some are tapered).

Tape path The complete path the tape takes as it leaves the supply reel, goes by the tape guides, erase head, head drum, CTL head, is driven by the capstan and pinch roller, and wound onto the take-up reel.

Tape pattern The magnetic pattern on the tape composed of the video tracks, control track and audio track.

Target The part of a vidicon camera tube where the image is formed.

Telecine A device used to change a motion picture into a TV signal. In essence a telecine consists of a film projector and video camera connected together for optimum picture quality and minimum flicker.

Tension servo A servo system that automatically applies a brake to the supply reel as needed to maintain constant tension along the tape between the supply and take-up reel spindles.

Test pattern A pattern used to check transmission and equipment performance. Test patterns are designed to check resolution, contrast, aspect ratio (ratio of width to height of picture), the presence or absence of scanning distortion, and other picture characteristics at a glance.

Threading *See* Loading.

Timer recording mechanism A system that automatically triggers a VTR or audio tape recorder to begin recording at a preset time.

Track The magnetic pattern left on the tape when a signal is recorded by a head. Video tape has separate video, audio and control tracks for each signal recorded.

Tracking adjustment When a tape recorded by one VTR having a CTL head out of position, is played back on another deck, the video heads may not trace the tape at exactly the correct point and mistracking will result. Rather than actually move the CTL head, this tracking error can be corrected by electronically delaying the CTL signal. This control is called tracking adjustment, and is not required in V2000 format, where DTF is used.

U-loading A loading system in which the tape passes around the head drum in a U shape.

U-loading is easy on the tape and ensures a smooth tape travel along with making possible instantaneous control of recording, playback, fast forward and rewind modes.

Unbalanced Two conductors are needed for transmitting an electronic signal. If they are not equal in relation to ground (if one is grounded), the system is unbalanced.

Unloading While loading refers to automatic wrapping of the tape around the head drum, unloading refers to putting the tape back into its original position in the cassette after recording or playback.

VHS format The world's most popular system, developed by JVC and marketed by most VCR manufacturers. Two new variants have emerged, VHS-C (Compact) and VHS-LP (long-play).

Vidicon A kind of camera tube. The vidicon is relatively simple in construction, compact, cheap, and easy to use. It is popular in many applications. A camera tube is a type of vacuum tube located within the video camera. It has the role of converting the image formed by the lens into an electronic signal.

Video From the Latin 'I see'.

Video head The electromagnetic heads in a VTR that record the video signal on the tape and pick up the signal during playback. In helical scan VTRs two identical heads contact the tape while rotating at high speed. A ferrite core is employed as the chip contacting the tape surface.

Viewfinder On a video camera the viewfinder is a small monitor (about 2.5 cm) that shows the picture shot by the camera lens. It is placed in such a way that it gives the operator the impression of actually looking through the lens so that he can easily adjust focus and aperture and follow subject movement.

V2000 A European-developed video format for home use. It enjoys the advantages of DTF (dynamic track following) and a flip-over cassette, both 'sides' of which can be used.

White balance setting An object will appear to change colour depending upon the colour temperature of the light illuminating it. Therefore, when shooting with a colour camera under varying light conditions it is necessary to decide on a colour reference on which the expression of other colours can be based. White balance concerns the process of establishing this reference.

Wipe A technique for changing from one picture (from one camera) to another (from a second camera). The change takes place at a specific boundary that moves across the screen horizontally, vertically, or diagonally until the new picture completely replaces the previous one.

Wow and flutter Wow and flutter are tape transport speed fluctuations that may cause a regularly occurring instability in the picture and a quivering or wavering effect in the sound during recording and playback. Longer cyclic fluctuations (below 3 Hz) are called wow; shorter cycles are called flutter (3–20 Hz).

Y/C Nomenclature for the luminance and chrominance signal processing circuits.

YH Full bandwidth luminance signal which is the B & W picture.

YL Low frequency luminance signal within a camera for combining with the low bandwidth colour components, i.e. R-YL and B-YL. YL is equivalent to green.

Zoom lens A lens with a continuously variable focal length. Once focused on a subject, the image remains in focus and the effective aperture remains the same even if the focal length (and therefore magnification) is changed.

INDEX